THE
DESIGN
AND
CREATION
OF
JEWELRY

ROBERT von NEUMANN

LONDON
SIR ISAAC PITMAN & SONS LTD

First published in Great Britain, 1962

Reprinted, 1965
Reprinted, 1967
Reprinted, 1968

SIR ISAAC PITMAN & SONS Ltd.
PITMAN HOUSE, PARKER STREET, KINGSWAY, LONDON, W.C.2
THE PITMAN PRESS, BATH
PITMAN HOUSE, BOUVERIE STREET, CARLTON, MELBOURNE
P.O. BOX 7721, JOHANNESBURG, TRANSVAAL
P.O. BOX 6038, PORTAL STREET, NAIROBI, KENYA

ASSOCIATED COMPANIES
PITMAN MEDICAL PUBLISHING COMPANY Ltd.
46 CHARLOTTE STREET, LONDON, W.1

PITMAN PUBLISHING CORPORATION
20 EAST 46TH STREET, NEW YORK, N.Y. 10017

SIR ISAAC PITMAN & SONS (CANADA) Ltd.
(INCORPORATING THE COMMERCIAL TEXT BOOK COMPANY)
PITMAN HOUSE, 381–383 CHURCH STREET, TORONTO

SBN: 273 43568 X

I Dedicate This Book to ALICE

PREFACE

The significant increase in both the creation and the use of jewelry in the past few decades restates a truth that is many thousands of years old: man needs personal adornment.

After inventing the tools for defense and food gathering, earliest man exercised his total ability in the creation of objects to be worn and to be beautiful to his eye. The discovery of metals started a tradition of transforming the raw materials of metal, stone, and glass into objects of richness and delight that continues to grow and expand to this day.

The technology of metal working has become complex. Though many of the tools of the jeweler today are identical to those in use two thousand years ago, a constantly expanding refinement of techniques and materials has resulted in a body of information almost as varied as a science. Where in the past an apprentice might have started to absorb the knowledge of a craft at a very early age, virtually all present-day craftsmen begin their concentrations only after years of general schooling. The difficulty in arriving at a total knowledge of working a material has resulted in many abbreviations of older, more time-consuming, techniques. However, even with the increased use of time and energy-saving devices, there is much that only painstaking experience can teach.

There have been several excellent books written on the subject of jewelry making. Each has attempted—as far as is possible—to establish a simulation of the apprentice-master relationship. This is, of course, tremendously difficult since to put into words clearly enough the act of doing is almost impossible. It would take volumes to describe every action of hand and mind used in the design and creation of jewelry.

In writing such a book, the author has attempted to describe only his personal experiences with the art of jewelry making. These experiences have occurred in the creation of his own work and in helping to solve the problems of his students. As far as possible, the author has avoided including information about which he has no personal knowledge, and in a field as varied and complex as jewelry making there are always avenues of expression and technique which have not been explored. Where these are touched upon, the author wishes only to excite interest in the reader to explore these directions as fully as possible and to come to independent conclusions.

The value of a book written on this basis is that, for those who have already had experience in jewelry making, there is much information to be used in making comparisons of technique and practice. For those who have not yet started, the information is functional without establishing restrictions of expression.

The creation of jewelry has unique aspects different from any other art form.

It has one function—to be decorative. The definition of decoration can and should be as personal as the imagination can be. Jewelry can be as freely experimental as any art form but, perhaps more than other art forms, it needs an underlying foundation of craftsmanship to be completely valid. Describing this combination of factors—the freedom to invent and explore coupled with a thorough knowledge of technical factors—has been the goal of the author in writing this book.

ROBERT VON NEUMANN

ACKNOWLEDGMENTS

Without the constant co-operation of my colleagues in the field of jewelry making and the yearly stimulation of vital and interested students, this book would not have been developed.

Special thanks to Alice Boatright, Irvin and Bonnie Burkee, Philip Fike, Earl Krentzin, John Paul Miller, Ronald Hayes Pearson, Robert Pierron, Christian F. Schmidt, and Prof. Elisabeth Treskow for allowing me to demonstrate the great variety and beauty of today's jewelry through photographs of their work. I wish also to give special thanks to Mrs. Stanley P. Wyatt for her kind help and advice in developing the manuscript.

ROBERT VON NEUMANN

FRONTISPIECE

1. Sterling silver pin with opal; Earl Krentzin
   Photo by Earl Krentzin
2. Brooch of gold granulation and precious opal; Elisabeth Treskow
   Photo by Elisabeth Treskow
3. Pendant in amaranth wood and sterling silver; Irvin Burkee
   Photo by Irvin and Bonnie Burkee
4. "Beetle" necklace, gold; John Paul Miller
   Photo by The Cleveland Museum of Art
5. Dinner ring, 14K white gold and diamond, 1956; Ronald Hayes Pearson
   Photo by Ronald Hayes Pearson

Except where otherwise stated in the text, all photographs and drawings are by the author. "The Princess," silver and gold pin, shown on the half title page preceding the title page of the book, is also by the author.

CONTENTS

THE
DESIGN
AND
CREATION
OF
JEWELRY

1 · MATERIALS AND TOOLS

THE METALS AND ALLOYS

The materials of jewelry making are quite varied. Traditionally, metals of many kinds, gem stones, wood, ivory, bone, and vitreous enamels have been used in the creation of personal ornament. The development of new materials—the plastics—and the increased technology of metals has resulted in an ever-increasing range of expressive possibilities.

Perhaps the basic material in jewelry making has always been metal. Almost all of the commonly available metals—platinum, gold, silver, copper, iron, and so forth—have been used in jewelry making at one time or another.

The rarity and the beauty of gold and silver especially have long been prized in jewelry the better to express the preciousness of fine design and workmanship.

Metals for jewelry making may be obtained in many shapes and forms. In addition, many have been developed into alloys, each having its special quality and use. The jewelry maker today may purchase a sheet of metal of desired dimension and thickness from a refiner, even specifying that it be soft or hard. Wire of a number of shapes and a great range of diameters may also be purchased, but many craftsmen prefer to draw wire to required dimensions when needed by using draw plates.

Metals may also be purchased as small bars or ingots to be used in casting or forging.

Before describing the important aspects of a number of metals used in jewelry making, the system of measurement should be explained. Nonferrous metals —those not containing iron or steel—are measured in the United States by the Brown and Sharpe gauge system. A sheet of metal or a section of wire is measured in its thickness by inserting it into an appropriate numbered slot in the gauge plate. (See Figs. 1 and 2.)

When ordering sheet or wire from a supplier, it is necessary to state all of the dimensions: *length, width,* and *thickness (gauge)*. When ordering by weight (precious metals are weighed by the troy weight system), it is still necessary to

Fig. 1

Fig. 2

state *width* and *gauge* or, if wire, *gauge* and *shape*.

Major Nonferrous Metals for Jewelry Making

SILVER

Melting Points $\begin{cases} \text{Fine silver} & 1761° \text{ F} \\ \text{Sterling silver} & 1640° \text{ F} \end{cases}$

For a number of excellent reasons silver is most often used for hand-wrought jewelry. Fine silver, the correct term for pure, unalloyed silver, is the whitest of all metals, has the greatest luster, and —next to fine gold—is the most malleable and ductile of all metals. It is so malleable that it may be beaten into thin sheets or *leaves* 0.00025 millimeter (mm.) thick. At this point silver readily transmits light.

Silver is so ductile that 1 gram (Gm.), a piece as large as a small pea, may be drawn out as a wire more than 1 mile long.

Hammering, bending, and compressing silver between steel rollers hardens it, but careful heating to the correct temperature quickly softens it once more. This last process is called *annealing* and will be described in Chapter 3.

Fine silver is too soft for most jewelry purposes except when used as a base for enameling. Because of its softness, other metals (usually copper in small amounts) have long been added to silver to form a stronger alloy. This alloy of copper and fine silver is called *sterling* silver if the proportions consist of 925 parts fine silver and 75 parts copper per thousand parts. Though other metals may be used, the above amount of copper has been found to give silver the necessary toughness without too much reducing its ductility and malleability. In addition, the copper allows silver to be *oxidized* in controlled ways—a process which often enriches the surface quality of silver objects.

The gauges of sterling silver illustrated in Fig. 3 show some of the practical uses for sheet form silver.

Sterling silver wire may be found in all standard gauges, though those shown in Fig. 4 are used most commonly by jewelry designer-craftsmen.

B & S Gauge			Weight per sq. inch troy ounces
12		for very heavy rings and forged bracelets	.443
14		for heavy rings and bracelets	.351
16		for average rings	.278
18		for lightweight pierced designs	.221
18			.221
20			.175
22		for brooches, earrings, beads, buttons, and bezels for settings, etc.	.139
24			.110
26			.087

Fig. 3

Courtesy, Handy & Harman

GOLD

Melting Points:
Range from 1500° F to 1945° F

Gold is a dense, lustrous yellow metal, the most malleable and ductile of all metals. In its pure (*fine*) state 1 Gm. may be drawn into a length of wire 2 miles long. It may be beaten into a sheet so thin that 1 ounce (oz.) may be spread over 300 square feet (sq. ft.).

Pure gold, like fine silver, is too soft for most practical purposes. It has thus been alloyed with a number of other metals to form alloys which vary considerably in color, hardness, malleability, and melting point.

Silver added to gold reduces the depth of the yellow color and forms a greenish alloy when used in larger amounts. Copper deepens the yellow of pure gold, making it both redder and harder.

The triple alloy of gold, copper, and silver is very malleable and close to the color of pure gold. Alloys containing platinum or palladium form the white golds often used in the setting of precious gems. White golds are usually harder and more durable than other alloys of gold and lend themselves well to the delicate, though strong, settings required for faceted stones.

An interesting alloy of 20% aluminum and 80% gold results in a purple alloy which proves to be somewhat brittle for working. Zinc and nickel are two other metals commonly alloyed with gold to create new characteristics. As in the case of sterling silver, there are legal proportions for gold which must be accurate before an article may be stamped with a *karat* value.

Pure or fine gold is considered to be 24 karats of fineness. Alloy golds may be 22, 20, 18, 14, 12, or 10 karats, or even less. For example, 18K gold is an alloy of 18 parts pure gold plus 6 parts of another metal; 12K gold is only half gold; and alloys below 10K cannot be stamped legally with the *karat* or quality stamp.

The finest and more expensive hand-wrought work in gold is usually of 18K quality. It has a somewhat richer quality and oxidizes less than other usable karats. But 14K golds are of good color and, being less expensive, are most often used in jewelry making.

Karat golds in general require more force in working than does sterling silver, but a craftsman easily adjusts to this difference.

Sheet and wire of karat golds may be purchased in the same forms as sterling or fine silver. It is important, however, to remember that gold is heavier than sterling silver. A silver ring reproduced in 14K gold might be 26% heavier and in 18K gold as much as 50% heavier. This could have great bearing on the size and design of work in gold, which is probably more successful when handled with lightness and delicacy.

COPPER

Melting Point: 1981° F

Copper has been used since 8000 B.C. Since it is often found in its pure state, since it is quite malleable and durable,

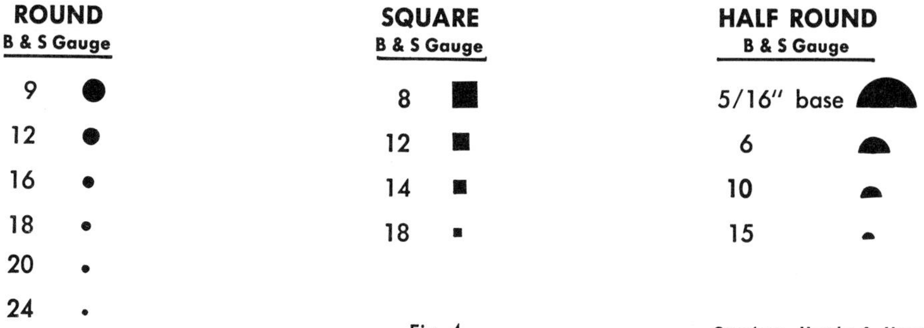

ROUND		SQUARE		HALF ROUND	
B & S Gauge		B & S Gauge		B & S Gauge	
9	●	8	■	5/16″ base	◗
12	●	12	■	6	◗
16	•	14	■	10	◗
18	•	18	▪	15	◗
20	•				
24	•				

Fig. 4   Courtesy, Handy & Harman

and because of its rich red color, early man found immediate decorative and functional uses for this metal. It is not often used for jewelry today because of its oxidizing tendencies, though, since vitreous enamels fuse to copper readily, there has been a recent increase in its use for enameled jewelry and other objects.

Copper's tendency toward rapid oxidation and sulfurization may be controlled by the addition of other metals to form alloys with a great range of characteristics. Many of these alloys may be used in jewelry making. The most important are:

BRASS

Melting Points:
Range from 930° F to 2075° F

Brass is an alloy of copper and zinc. The color is bright yellow and combines with the white of silver and the red of copper quite decoratively. Even though the melting point of so-called standard brass is around 1800° F, it has a tendency to collapse into silver, when soldered to it, at a much lower temperature.

BRONZE

Melting Points:
Range from 572° F to 1926° F

Basically an alloy of copper and tin, this versatile metal can be made to be as soft as pure copper or as hard as some steels. In color it ranges from a warm red-yellow through gold to dark brown.

Aluminum bronzes—alloys of copper and aluminum plus small amounts of other nonferrous metals—have high tensile strength and clean casting qualities, and besides are acid- and oxide-resistant. Melting points of these alloys range from 1130° F to 1926° F.

Phosphor bronzes—alloys of copper, tin, phosphorus, and zinc—are true bronzes to which small amounts of phosphorus have been added as deoxidizers

and strengtheners. They are very hard, springy alloys.

NICKEL SILVER

Melting Point: 1959° F

Nickel silver—or German silver, as it is often called—contains no silver at all, consisting of approximately 60% copper, 20% nickel, and 20% zinc.

Nickel silver is a strong, ductile alloy resistant to oxidation even at high temperatures. It has a slightly yellowish gray quality which makes it less rich in color than silver.

MONEL METAL

This is an alloy of 29% copper, 68½% nickel, 1% iron, 1% manganese, silicon, sulfur, and carbon.

Monel metal is a rather dark gray alloy of considerable strength and great oxidation resistance.

PEWTER

Melting Points: From 500° F

Once an alloy of lead and tin alone, more modern alloys of pewter consist of a combination of copper, tin, and antimony. Pewter is quite soft in most forms and is easily formed by repoussé and smithing techniques. Its low melting point requires the use of soft solders which lack strength and precision for fine work.

There is one major difficulty in combining metals such as silver and copper by soldering. Hard solder used in much jewelry making is a light-colored alloy of silver, copper, zinc, and cadmium. Since it is light in color, a misplaced piece of solder or an overabundance of solder may easily form an unsightly blemish on a darker metal. Consequently, greater than usual care must be exercised in placing solder accurately and in heating carefully during soldering.

Copper, brass, and bronze may be given a high luster by buffing, a warm matte tone by brushing, or a rich patina of black, brown, green, or green-blue by the application of various chemicals. The application of color to metal is described in Chapter 2.

Ferrous and Other Metals

IRON, STEEL, AND ALUMINUM

Though these metals may be used in jewelry making, they often present problems in work techniques which make them less functional than the previously described metals.

Iron and steel are difficult to combine by soldering in a precise or delicate manner. Much stronger joins are made in these metals by welding or brazing, both techniques being somewhat coarse for jewelry purposes.

Aluminum, in addition to having a rather unpleasant lack of weight and substance, is even more difficult to solder well. Recent developments in solder alloys and fluxes for aluminum have resulted in improved strength and durability in joins, but again do not lend themselves to more precise jewelry soldering needs.

Often the inherent restrictions in metals and other materials force the imaginative designer-craftsman into new paths of experimentation. Therefore, all materials should be examined as potential media and none discarded merely because of traditional disinterest.

ENAMELS

Vitreous enamel, a form of fused glass, has long been combined with metal as a decorative art form. Like glass, it is hard, brilliant, and permanent. Colors have not lost intensity and richness in the thousands of years since first having been fused to metal.

As early as 500 B.C. the Greeks had already known much of the technology and use of enamels, and this knowledge spread north and east until it found fertile ground in Europe, China, and Japan, where enameling eventually developed as an important art form in itself.

Enamel is composed of a basic flux or *frit*, which is colorless, and various metallic oxides which give it color or opacity.

By combining a number of oxides in a great variety of proportions, hundreds of colors have now been developed. In addition, clear or translucent colored enamels vary in hue and intensity depending on the thickness of a layer as well as on the metal to which they have been fused by heat. Today enamels are being used not only on all of the traditional metals (such as gold, silver, copper, and alloys of these metals), but also on iron, steel, and aluminum as well.

Enamels may be purchased already ground and graded to specific particle sizes. These graded sizes are used in a number of enamel applications, such as painting, dipping, spraying, inlaying, screening, and sifting.

Enamels may also be purchased in lump form, after which they may be ground in an agate or mullite mortar to the particle size desired. Enamels tend to decompose slowly when stored as ground particles but last much longer in lump form.

Enamels may be transparent, translucent, or opaque. Combinations of colors having these qualities may be used to create depth and variety in design.

Enamel surfaces after fusion to metal may range from glassy brightness to matte softness.

There are a number of traditional application techniques readily adaptable to the experimental approach of contemporary jewelry. These techniques, both historical and contemporary, are described in Chapter 4.

GEMS—PRECIOUS AND SEMIPRECIOUS

Gems have always been considered an important element in jewelry. The rarity of a richly colored stone, its brilliance when polished, and its effective accenting of a form in metal have all combined to make a gem precious.

In earliest times the most colorful stones which could be worked easily were of greatest value. In Egypt, where jewelry performed an unusually important function in society, gems such as turquoise, lapis lazuli, carnelian, agate, and coral were used lavishly in necklaces, head ornaments, and bracelets. As the skill of gem cutting—*lapidary*—developed, harder materials such as quartz, amethyst, ruby, sapphire, and diamond were introduced into jewelry.

Today we have a range of hundreds of gem materials which may be purchased in the form of shaped and polished gems or which a craftsman may shape to his own purpose, beginning with raw materials.

The setting of gems in metal and the basic lapidary techniques are described in Chapter 4.

WOOD

Though not widely used in historic jewelry, in its variety of color and grain wood is finding ever greater application in contemporary jewelry design. It has been used as a frame or background for metal forms, as an alternation with metal in repetitive forms, or as a sculptural element independent of other materials. It may also be inlaid into metal or have metal imbedded into its surface.

The tropics of the world supply the craftsman with a number of rare and handsome woods which may be purchased in small amounts from firms dealing in materials for fine cabinet work. Such woods as ebony, cocobolo, zebra wood, snake wood, amaranth, and rosewood are all beautiful enough and durable enough to be used as a precious material.

Some of our native hardwoods—birch, black walnut, cherry, and oak—are often locally available. All hardwoods, those of a close and compact grain, may be worked with virtually the same tools used in metal work.

The processes of cutting, forming, fastening, and finishing wood for jewelry are described in Chapter 4.

PLASTICS

Of all the materials available to the jewelry designer-craftsman today, none contains as much potential as the plastics. Almost unknown before the twentieth century, they form for today's craftsman a real challenge in use and interpretation.

Since plastics may be used as solids, semisolids, or liquids, the range of interpretative possibilities is unlimited.

Using standard metal-working tools and equipment, the craftsman may saw, file, carve, grind, drill, and polish plastics such as lucite or plexiglas and nylon. He may cast specific shapes and imbed or enclose decorative materials of other kinds in the slow-hardening liquid polyester plastics.

Plastics are not only manufactured in a range of colors but they may also be dyed by the craftsman himself.

Plastic materials are so versatile that the great danger lies in the conscious simulation of other, perhaps more difficult to work, materials. True honesty of workmanship exploits a material only in directions that are basic and honest to that material and to create mineral or wood imitations in plastic, for example, is a fraud and not deserving of time or effort.

Techniques of plastic uses are described in Chapter 4.

ADDITIONAL MATERIALS

History provides us with a number of other materials which, with the freedom of expression characteristic to contemporary design, may be interpreted in fresh and exciting ways.

The ivory of elephant, walrus, and rhinoceros, bones and teeth of many sorts, a great variety of shells—all are examples of the original materials used by early man in his search for decorative expression. These materials may be carved, drilled, sawed, filed, and polished by use of metal-working tools and techniques. Often the grain is interesting enough to be exploited as an important design element.

Seeds, bamboo and other reeds, and even insects have been used by early or primitive craftsmen where metal or stone was not available or not known. Seeds may be drilled, set as gems, or cut into sections for repetition of shape. Reed and bamboo, because of their tubular rigidity and varied surface textures and colors, may be used in such jewelry as necklaces and bracelets in countless ways.

Insects, whole or in part, may be fused into plastic and used as central or repeated motifs. The hard bodies of domestic and tropical beetles, colorful and intriguing in shape, may be set as gems are set.

The materials mentioned are not all that may be used in jewelry making by any means. In the search for the better expression of a creative need, an imaginative designer-craftsman will discover for himself countless new and fresh applications. In this sense he is a true creator.

FINDINGS

Since the materials of jewelry must be adapted to some means of wearing them, the range of mechanical fittings—*findings*—must also be described in this section. The means of applying them to jewelry will be described specifically at other points in this text.

A great variety of shapes and qualities in findings is available to the craftsman and it often requires discrimination to decide which might be best in a design.

There are several degrees of quality in findings and often the same shape and function may be found in both cheap and expensive materials. The least expensive findings are made of yellow or white metal-plated brass or nickel. These do not wear well, often become soft in soldering, and should not be used in fine jewelry. For a small increase in cost it is best to use sterling silver findings for silver jewelry and karat gold findings for gold jewelry.

Findings are manufactured for virtually every purpose and are often quite well designed. Even so, many craftsmen prefer to design and construct a finding for a specific piece of jewelry, in the belief that it will be better integrated with the piece.

For Earrings

1. Standard screw type.
2. Screw type with link.
3. Screw type with dome.
4. Clip.
5. Wire with link, for pierced ears.

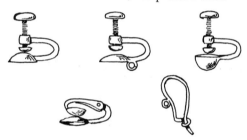

For Pins and Brooches

1. Side opening catch, joint and pinstem with fixed rivet.
2. Side opening catch, joint with soft-soldering patch. Pinstem with loose rivet.

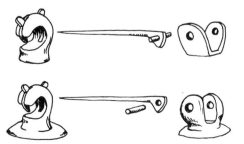

For Necklaces and Bracelets

1. Spring ring.
2. Round jump ring.
3. Oval jump ring.
4. Foldover catch.
5. Box catch.
6. Chains: curb link, round link, fancy, fancy.

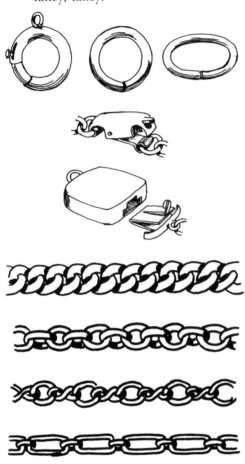

For Tie Tacks

1. Tie tack with holding spur.

For Cuff Links

1. Cuff link back with separate rivet and joint.

TOOLS AND EQUIPMENT FOR THE WORKSHOP

It is interesting that creative man had, at an early time, designed the tools of his arts so well that many have not changed in shape and function to this day. Of course modern technology has improved the quality of the tool and mass production has made a tremendous variety available to the craftsman who had to make his own in past ages, but the functional aspect has changed very little.

The tools of the designer-craftsman need not be complex or great in number.

The Indian silversmiths and jewelers of the American Southwest practice great economy in their work. The hardware store supplies a ball-peen hammer, a few large and small files, and sandpaper. A blow torch supplies heat and, lately, an electric motor speeds the polishing operation.

Though the work is often quite simple in concept, it reflects the fact that a workshop full of expensive gadgets is not of primary importance in jewelry making. What *is* important is that the right tools for the work at hand are chosen with care.

As in all purchases, it is a foolish economy to buy the least expensive of several makes of a tool. Cheap pliers mar the materials they work on and often break under normal pressure. Files which are poorly made are rough and uneven in cutting surface. Saw frames can break under the tension necessary to insert a blade.

A sense of confidence in a tool is a value which can reflect itself in the finished object. In addition there is great aesthetic pleasure in handling a well-designed and beautifully made tool and it becomes a pleasure to maintain it in good condition.

In the following lists of tools, four categories are formed.

The first lists basic equipment without which the designer-craftsman is hampered in his explorations of techniques and materials.

The second list adds additional small tools which, though not of primary necessity in the beginning, later become necessary to the organization of a complete workshop.

The third list includes the items of large equipment basic to the workshop. In some cases, their greatest contribution comes in the saving of time. One can be sure, contrary to a romantic nineteenth-century philosophy, that the level-headed craftsmen of earlier times would have found these economies worth while had they been available.

The fourth list contains additional large equipment which might complete a well-organized workshop. In some cases the high cost of these tools may prohibit their purchase by the individual craftsman. It is worth while investigating local recreation and school workshop facilities. Often these areas make machinery and space available to interested individuals or groups.

The catalogs and price lists of the better jewelry tool and supply houses often illustrate a complete range of tools and materials available. The names and addresses of the major supply companies are listed on pages 209-210. Local hardware stores are also good sources except for the highly specialized tools of the craft.

List 1—Basic Small Tools

For Sawing, Cutting, and Filing

1. Jeweler's saw frame, 5″, adjustable

2. Jeweler's saw blades:
    Nos. 0 and 1 for average work
    Nos. 2/0 and 3/0 for fine work

3. Assorted needle files, No. 1 or No. 2 cut, 5½″ length
    Round
    Half-round
    Crossing
    Barrette
    Square
    Knife

4. Large hand files, 6″ cutting length, No. 1 or No. 2 cut, wooden handles
    Flat
    Half-round
    Round
    Triangle

5. Riffler files
    No. 7, flat and curved
    No. 10, spoon: half-round and curved
    No. 17, pointed, half-round
    No. 5, knife

6. File brush

7. Bench pin:
    Jeweler's bench pin
    or
    V-board and clamp

8. Jeweler's shears:
    Plate shears—scissors handle
    or
    Brown's jeweler's shears

9. Diagonal nippers

10. Hand drill

11. Drill bits—graduated sizes

12. Ring clamp

13. Beeswax

14. Center punches and scribes

15. Small smooth-jawed bench vise

16. Scotch stones

For Forming

1. Pliers with smooth, polished jaws:
   Round-nose
   Chain
   Flat-nose
   Half-round
   Rivet-setting
2. Chasing hammer with convex polished head
3. Scraper, hollow
4. Burnisher, straight, curved, with narrow 2" blade
5. Lead block
6. Steel block, polished
7. Hardwood block:
   Solid maple
   Scored maple
8. Wood mallet, curved and flat end
9. Ring mandrel with graduated sizes

For Decorating

1. Chasing tools
2. Potassium sulfide (liver of sulfur)

For Soldering

1. Tweezers, smooth and polished:
   Fine-pointed
   Locking
2. Small brushes for flux and solder application
3. Binding wire
4. Asbestos (or transite) sheet
5. Charcoal or asbestos block or an asbestos ring
6. Solder:
   Silver solder in sheets, strips, or wire
   Gold solder in sheets, strips, or wire
   Lead solder wire
7. Silver solder flux, either paste or liquid
8. Lead solder flux
9. Pickle pan or jar with cover:
   Copper
   or
   Glass
   or
   Stoneware
10. Pickle tongs

11. Pointer
12. Yellow ocher to protect solder joins
13. Soldering unit:
    Gas-air
    or
    Acetylene
    or
    Mouth blow pipe and gas
    or
    Propane gas in self-contained unit
14. Sulfuric and nitric acid
    or
    Sparex pickling compound
15. Flint striker

For Buffing and Polishing

1. Emery paper, at least two sizes, Nos. 1 and 3/0
2. Buffing sticks and boxwood pegs
3. Rouge cloth
4. Abrasives:
   Tripoli
   Pumice
   Rouge

Miscellaneous Tools

1. Steel rule
2. Scribe
3. Compass

List 2—Supplementary Small Tools

For Sawing and Cutting

1. Jeweler's saw frame, 8", adjustable
2. Scorer, small

For Forming

1. Dapping block and dapping punches
2. Steel bending block
3. Bezel mandrels:
   Oval
   Square
   Round
4. Small stakes
5. Large stakes and holder
6. Smithing hammers

7. Pitch bowl
8. Gem pusher, square, round
9. Beading tools
10. Seating drills and burrs
11. Plastic mallet
12. Rubber mallet
13. Horn mallet
14. Bracelet mandrel

For Decorating

1. Engraving burins
2. Arkansas stone
3. Oilstone
4. Sealing wax
5. Additional chasing tools
6. Matting tools
7. Acid resist for etching
8. Pyrex glass tray

For Soldering

1. Locking tweezers with holding stand
2. Asbestos ring-soldering mandrel

For Buffing and Polishing

1. Buffing and polishing wheels of cotton, muslin, wool, etc.:
   Stitched
   Unstitched
   Lead center
   Goblet
2. Felt buffing wheels
3. Felt and emery ring-buffing mandrels
4. Flexible-shaft machine
5. Buffing and grinding equipment for the flexible-shaft machine

Miscellaneous

1. Gem holder
2. Draw plate: round hole, square hole, half round hole
3. Drawing tongs
4. Gauge plate
5. Set of ring sizes
6. Jeweler's loupe
7. Stamps:
   Hallmark (maker's name or sign)
   Sterling

14K and 18K (for gold)
Assorted numbers
Assorted letters
8. Slide caliper—inches and millimeters
9. Hand reamer
10. Washout brushes
11. Jeweler's screw plate for threading holes and wire

List 3—Basic Large Tools

For Sawing and Cutting

1. Bench shears, 4″ blade
2. Small drill press with electric motor

For Forming

1. Additional stakes
2. Additional mandrels for bracelets, etc.
3. Large vise
4. Centrifugal casting equipment:
   Machine
   Casting flasks
   Sprue formers
   Sprue pins
   Crucibles
   Crucible tongs
   Casting wax—bars, sheet, and wire
   Casting flux

For Decorating

1. Enameling kiln and equipment:
   Trivets
   Sieves
   Spatulas
   Carborundum stones
   Cloisonné wire
   Gold and silver leaf
   Assorted ground and lump enamels
2. Plating machine and equipment

For Soldering

1. Pumice pans, rotating

For Buffing and Polishing

1. Polishing machine with hoods and dust collection

1. Jig saw, electric
2. Belt sanding machine
3. Rolling mill
4. Lapidary equipment:
   Cabochon unit with saw, grinding wheels, sanding and polishing discs
   Facet-cutting machine
5. Jeweler's workbench
   The ideal workbench is designed specifically for jewelry making, but an ordinary table or desk will do if it is sturdy enough. The workbench should be located near water, good light, and ventilation. Soldering and pickling may be carried out at the bench unless large pieces are made. In that case, a fire- and acid-proof work area should be designed containing adequate fume-venting facilities.

ORGANIZATION OF THE WORKSHOP

Though it is possible to make jewelry in virtually any space large enough to hold a table or a desk, it is far better to have space designed specially for this activity. An ideal room has, in addition to good traffic space, the following assets:

Good Light

This may be incandescent light in the form of adjustable lamps for each work area (i.e., construction, soldering, pickling, buffing, etc.). Fluorescent fixtures are adequate but they place a greater strain on the eyes when one is working with small objects.

Water

A convenient source of hot and cold water is almost imperative for washing out work after pickling in acid and after buffing and polishing.

Adequate Electricity

Since much equipment is motor-driven or uses heavy current loads in heating elements, an adequate 220-volt line should be laid in. Enough wall or baseboard outlets should be supplied so that overloads on extension cords are prevented.

Ventilation

At best, a hood should be placed over a common soldering, annealing, and pickling area. This should have a fan strong enough to pull out dangerous fumes and gases quickly. Lacking this, window ventilation should be adequate in amount and placement.

Proper Wall, Ceiling, and Floor Surfaces

Since much metal work requires considerable hammering, it is wise to soundproof walls and ceilings as well as possible. If cleanliness is to be considered, these surfaces should be washable as well as light in color in order to increase total lightness in the room.

The floor should stand up under the moving of heavy objects and the occasional spilling of water or acids. Painted concrete or industrial tile is adequate.

Storage Space

Built-in or spatial cabinets for tool, material, and chemical storage should be designed for ease in use, maintenance, and safety. It might be well to provide locks where children and others might create a safety problem. (See Fig. 5.)

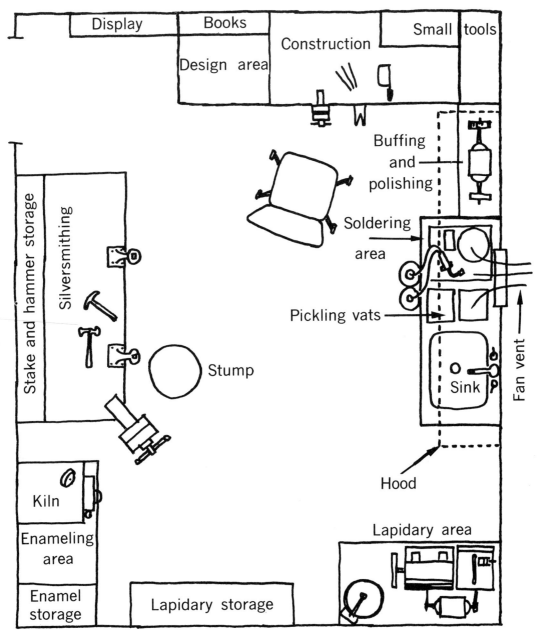

Display   Books   Construction   Small tools

Design area

Buffing and polishing

Stake and hammer storage

Silversmithing

Soldering area

Pickling vats

Stump

Sink

Fan vent

Hood

Kiln

Enameling area

Lapidary area

Enamel storage

Lapidary storage

Fig. 5

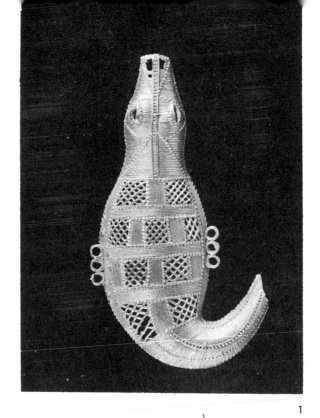

1

The jewelry and other small metal objects of Central and South America, Africa, Asia, and Europe reflect not only the religious and the social values of these ornaments, but also a great delight in the forms of living things. In almost all instances, the shapes of animals, birds, fish, insects, and man himself have been enriched, reorganized, and imaginatively interpreted.

The skill with which artisans of early cultures fabricated ornaments in gold, silver, and jewels is impressive when one considers the simple tools and materials known to them. Though most of the very intricate work was first modeled in wax and then cast in the *cire-perdue* process, many cultures knew and practiced the intricacies of soldering, forming, and stone cutting.

The Portuguese in West Africa and the Spaniards in Central and South America wrote glowing accounts of the delicacy and the richness of the ornaments found there. It was difficult for them to believe that people they considered simple and savage in other respects could have produced work of such sophistication that it rivaled some of the best of the Renaissance.

For the contemporary artist-designer, work of these cultures is significant in the great variety of form interpretation—from the most naturalistic to the utmost in expressionism.

1. Gold crocodile. Lost wax casting. Ivory Coast, Africa.
   The Cleveland Museum of Art, John L. Severance Collection
2. Christ medallion. From the Guelph Treasure, German (Frankish),
   8th century, gold and cloisonné enamel.
   The Cleveland Museum of Art, J. H. Wade Collection
3. Gold mask, embossed. Mochica Culture, Peru.
   The Cleveland Museum of Art, Mr. and Mrs. Henry Humphreys Memorial
4. Gold figure. Lost wax casting. Chibcha Culture, Colombia.
   The Cleveland Museum of Art, Mr. and Mrs. Henry Humphreys Memorial

1

2   3   4   5

Never in history has the art of the jeweler assumed such importance to society as it did during the 14th to the 18th centuries in Europe. Both men and women of the aristocracy and the wealthy merchant class adorned themselves with garlands of gold chains encrusted with rubies, pearls, and sapphires. Each hand wore several rings—often several on each finger. Medallions of gold and jewels were sewn to clothing of rich velvets, silks, and furs, and the need for displaying ever new, ever more impressive jewels had gold- and silversmiths by the hundreds working to their highest capacity.

During the 15th and 16th centuries especially, the skills of *cloisonné, champlevé, plique-à-jour,* and *grisaille* enameling, as decorative enrichments of jewelry, developed to an extremely high level. The cutting, polishing, and setting of gems became a major industry as well as establishing shapes and uses still popular today.

1. Necklace with a pendent sphinx in gold, gems, and enamel. In the style of Benvenuto Cellini. Italian, 16th century. The National Gallery of Art, Washington, D. C., Widener Collection, 1942

2. Pendant representing Europa and the Bull in gold, gems, baroque pearl, and enamel. In the style of Benvenuto Cellini. Italian, 16th century.
The National Gallery of Art, Washington, D. C., Widener Collection, 1942

3. Pendant, representing a centaur in gold, gems, baroque pearls, and enamel. Italian School, 16th century. The National Gallery of Art, Washington, D. C., Widener Collection, 1942

4. Pendant representing a triton in gold, gems, baroque pearls, and enamel. Italian School, 16th century.
The National Gallery of Art, Washington, D. C., Widener Collection, 1942

5. Pendant representing a mermaid, in gold, gems, baroque pearls and enamel. In the style of Benvenuto Cellini. Italian, 16th century.
The National Gallery of Art, Washington, D. C., Widener Collection, 1942

6. A goldsmith's shop, circa 1576. Copper engraving by Delaune. The Bettmann Archive

6

2 · BASIC TECHNIQUES

Each craftsman, having worked with his craft over a period of time, finds that it is always possible to modify—even to change radically—the methods used by other craftsmen of the past and of the present. The steps of construction in metal and other materials outlined in this and following chapters are the result of experience in teaching and personal jewelry work, and describe *one* way of doing—never the only way of doing.

Perhaps, at times, the description of a process may seem unnecessarily complex. In attempting to establish a more complete understanding of the logic of approach, detailed analysis is not only necessary but also justification for such a book as this. To actually see—and thus to participate quite closely—in the development of a process is the most ideal learning situation. Descriptive words must always fall short of such participation, but when developed in deep detail this unfortunate gap may be somewhat reduced.

Again—the description of a given approach to a problem, whether of design or of construction, represents only one craftsman's experience and is not to be taken as ritual or aesthetic law.

DRAWING AND TRANSFERRING A DESIGN TO METAL

When making the preliminary sketches for a piece of jewelry, always anticipate problems of construction. Keep in mind, while designing, such factors as ultimate size, weight, and strength. Know your materials well enough to avoid designing forms which would not survive construction or wear.

If a metal to be used is rather soft and easily bent, avoid designing shapes which project away from a supporting surface. *This is an important factor in using wire.*

The placement and types of findings should be planned for at this stage so that they will function well, will be safe when worn, and will not interfere with the unity of the design.

Preliminary sketches may be complex in showing several views of a piece. They may indicate textures accurately and plan areas to be oxidized. They may be rendered accurately to show dimensions and surface reflections. On the other hand, a sketch may be no more than a simple linear indication of shapes and forms.

One of the real pleasures in designing freely for personal satisfaction comes in allowing the tools, materials—even chance—to dictate some of the decisions. Often a texture or shape change comes about during construction which gives added effectiveness to the original design concept. Remain flexible in attitude. In avoiding an arbitrary narrowness in designing one can make of the purely mechanical—and often time-consuming—construction of jewelry a constantly interesting and challenging experience.

Once the sketch has been made, there are several ways in which its parts may be transferred to the working metal:

1. Many designer-craftsmen re-sketch directly on the metal. A freely evolved design might even benefit by having been interpreted just once more!

2. The back of the paper design might be blacked in with a soft lead pencil and the design transferred by drawing over the lines when placed over the metal. If the design is drawn on tracing paper, it is possible to see the shape and the limits of the sheet of metal beneath it. This allows an economical placement of the design and reduces waste.

3. A similar transfer technique is to place carbon paper between the sketch and the metal. This may make a rather heavy dark line which might blur or hide small details in the sketch.

4. A highly professional but more time-consuming technique consists of cleaning the surface of the metal to free it of grease film. A thin layer of Chinese white tempera is painted on and allowed to dry. A carbon or pencil-rubbed impression then shows up very well after transfer.

After any of the foregoing procedures, it is wise to use a pointed scribe to lightly scratch over the pencil or carbon lines, for during sawing the latter might be rubbed off in handling. Make the scratches light since chance and a change

of mind might dictate a new direction later, at which time a deep scratch would be difficult to remove.

5. If the craftsman wishes to remain precise and accurate, the sketch may be cut out of the paper and rubber-cemented directly to the metal. For a good join allow a layer of cement to become almost dry on both paper and metal before joining them. In this way the saw will cut through both the paper design and the metal at the same time.

SAWING AND PIERCING

Materials

    Jeweler's saw frame
    Jeweler's saw blades
    Beeswax
    Bench pin, "V" clamp or vise

A skillful craftsman with a good saw can approximate with a saw cut what can be done with a pencil line. By following a few basic rules, and with practice, it is possible to cut out any shape desired. The saw may be used to cut out simple basic forms or to create the most complex linear pattern.

In selecting a saw for metal sawing, make sure that the clamps holding the blade ends fit together smoothly. Be sure also that the frame is adjustable so that it may be easily lengthened or shortened. Select a saw frame that is 3″ to 4″ deep for average cutting. A deeper saw frame, though more versatile, is much less easily controlled while sawing because of the poor weight balance. If necessary, purchase an extra saw frame 6″ or more deep, from the blade to the back of the frame, for cutting deeply into a sheet of metal. (See Fig. 6.)

The blades to be used should be the best available. Generally, those made in Switzerland and Germany give the best quality for the cost. Blades vary in size; an "8/0" is very fine and a "14" very

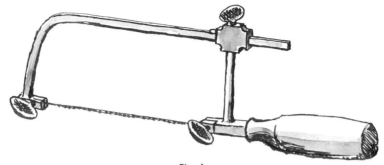

Fig. 6

coarse. For general work, a "0" or a "1" will do very well. A "2/0" or a "3/0" is useful for delicate linear work but they break easily if not controlled at all times.

Beeswax is used to lubricate the blade, thus reducing wear and speeding the sawing operation, but, if applied too often, beeswax tends to clog the saw teeth and thus reduce the efficiency of sawing. Excess wax may be brushed out of saw teeth with a bristle brush.

Many jeweler's benches come equipped with a slot holding a hardwood pin into which a "V" has been cut. This *bench pin* forms a support for the sheet of metal while you are sawing horizontally. (See Fig. 7.)

For workbenches without the above equipment, a *"V" board and clamp* may be purchased for the same purpose or a hardwood board might be cut to shape and fastened with a "C" clamp. The circle at the end of the "V" cut enables delicate sawing of small pieces while furnishing support at the necessary points. (See Fig. 8.)

A small bench or machinist's vise may

also be used if precautions are taken to protect the sawed metal from excessive jaw pressure or marring due to rough jaw surfaces. Many craftsmen line the vise jaw faces with smooth hardwood, leather, or sheet cork.

Attaching the Blade

Attaching the blade properly is important! Being made of fine tool steel, jeweler's blades are brittle and thin and should be handled accordingly.

Step 1. Loosen the jaw nuts on both ends of the frame.

Step 2. Insert the blade all the way into the top jaw nut. Make sure that the saw teeth face *away* from the frame back and angle *toward* the handle. Tighten the top jaw nut.

Step 3. Brace the end of the saw frame against the bench so that the frame back hangs down and the handle faces you. (See Fig. 9.)

Step 4. Make sure that the loose end of the saw blade almost but not quite reaches the bottom jaw nut. This added length is necessary when com-

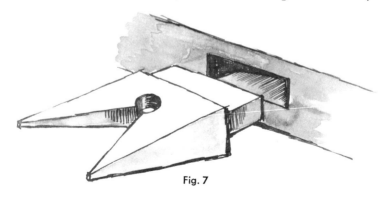

Fig. 7

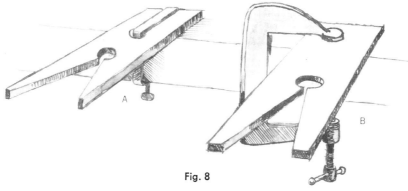

Fig. 8

pressing the frame. Compress the frame by pressing it against the bench and, while compressed, insert the loose blade end, tighten it into place, and release pressure slowly, for if pressure is released too quickly the sudden strain may snap the blade. The blade should be rigid and give a "pinging" sound when plucked.

Starting the Saw Cut

When starting a cut from the edge of a sheet of metal, the blade tends to stick or skid from place to place. This may be avoided by starting a groove in the edge by a few upward strokes of the blade. The actual cutting takes place on the down stroke, but a few strokes in the other direction help at the start.

Sawing

There are basically three aspects of sawing that should be mastered to prevent excessive blade breakage. The first rule is that *the blade must be kept perpendicular to the sawing surface of the metal at all times.* If the blade is at an angle, it becomes pinched when making a curved cut or when it is withdrawn from an incomplete cut. Pinching breaks the brittle and tightly strung blade.

The second rule is to *avoid excessive forward or downward pressure while moving the saw.* The weight of the hand is enough to draw the blade through anything but the thickest and toughest metal. In addition, there is a natural forward motion to the saw stroke so it is unnecessary to push the blade forward

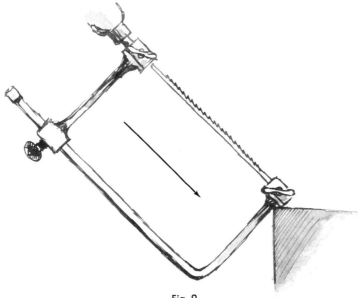

Fig. 9

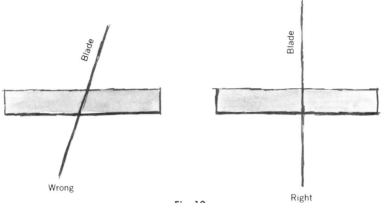

Wrong

Right

Fig. 10

with each stroke. Take it easy! Don't saw too fast—you lose control—and always use as many of the cutting teeth of the blade as possible with each stroke.

The third thing to keep in mind is to *use no force when backing out of a cut.* The blade must come out the way it went in. Pinching at this point breaks many blades. If the saw blade is deep in an intricate passage of lines, it might be best simply to release the bottom end of the blade, pull it out of the metal, and reset it again.

It is possible to saw the most intricate arabesques as well as sharp and precise angular shapes. When an angle of any degree is desired, one merely saws forward on one leg of the angle to the apex or point of turning, pauses at that point while continuing to saw up and down without forward motion, and slowly turns the entire saw frame to the desired degree of angle. Once the new angle has been reached, the saw may again travel forward.

In sawing curves of any degree the same action takes place. The saw frame is slowly turned in the desired direction while sawing up and down and forward. Some craftsmen keep the saw in one forward position while moving the metal about. Others move both saw and metal as necessary. One soon evolves the most natural technique for the job.

Piercing

When it is necessary to cut out a hole or a negative shape from the interior of a sheet of metal, a somewhat different start is made. After the sketch is transferred to the metal, a small indentation is punched at some appropriate point on the inside of that shape. This dent acts as a start for a small (No. 60) drill bit which drills the hole through which the saw blade is inserted.

Important! When using a punch always place the metal on a *flat metal* surface, such as an anvil or a bench block. If the punching takes place over wood or some other soft surface, the metal around the indentation will be bent and depressed. This sort of blemish is difficult to remove.

After the saw blade is inserted through the drilled hole, the blade is set again as before and sawing may proceed. When the saw arrives at the drilled hole again, the blade is released and the cutout metal removed. (See Fig. 11.)

In all sawing it is wise to saw on the

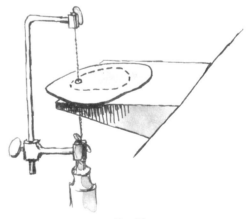

Fig. 11

waste side of a sketch line so that if the saw cuts irregularly the mistakes may be remedied by filing and do not infringe on the planned proportion of the design.

Blade Breakage

Often, when a blade breaks, fairly large sections of the cutting teeth remain intact. By making the saw frame shorter, these fragments may still be used. Since the toothed portion is more brittle than the ends, the blade is liable to break more easily the second time so that it will be necessary to be more careful in sawing.

Saw blades do wear out. When the teeth are dulled to the point where extra

Filing

There is always a correct file to use for the job at hand. Some files are limited to one kind of work while others may be used interchangeably. In all filing an efficient stroke of the file surface over the metal is important. To file incorrectly wastes time and energy and often causes more trouble than it solves.

Basically, an efficient cutting stroke consists of filing from the tip of the file to the handle. (See Fig. 12.) Many craftsmen develop the habit of filing from tip to handle and lifting the file from the metal surface at the end of each stroke. Others leave the file in contact but allow it to slide *lightly* back to the tip. Pres-

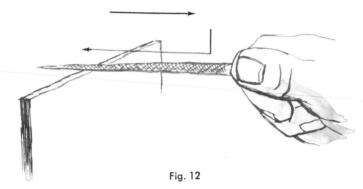

Fig. 12

force must be used in cutting, it is best to replace the blade. Extra pressure decreases control, so nothing is gained by this economy.

FILING, SCRAPING, STONING, AND BURNISHING

Materials

Large hand files: round, half-round, flat, all No. 2 cut
Needle files: round, half-round, triangular, bird-tongue, knife, flat, slitting, crochet, joint-finishing, barrette, equaling, square
Riffler files
Hollow scraper
Burnisher, curved or straight
Scotch stone

sure is then again applied on the cutting stroke.

Since filing is used most often for refining or correcting sawed edges, an economy of motion should be developed to do this well and quickly.

To file along an edge in line with the edge causes the file to slip off to the side. To file away high spots only where they occur causes too many depressions which also must be removed. It is best to file at a diagonal to the filing surface with a long, even, sliding stroke. If this eventually causes rough parallel grooves in the filing surface, they may be removed by filing from a new tangent. This slanted sliding stroke should cover the greatest distance possible and may be used on either convex or concave surfaces. (See Fig. 13.)

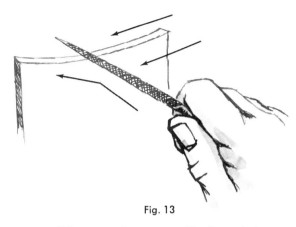

Fig. 13

Of course, the correct file shape is important here. For concave edges a half-round, bird-tongue, or round file should be used. For convex edges any of the several flat-sided files will work well.

Using the sliding stroke efficiently results in filing many high spots with each stroke and rapidly arriving at a clean even edge or bevel.

While being filed, work may be held in the hand, in a ring clamp, or in a vise.

When holding metal by hand alone, it is necessary to brace it against the workbench edge or some part of the bench pin. It is possible to design a wooden projection fastened to the bench that will act as a firm, easily approached support.

At times the sections of metal to be filed are either too small or too fragile to be held by hand. A ring clamp makes an ideal and safe holder for this purpose. The work should be placed far enough into the jaws of the ring clamp to prevent bending or breaking with the pressure of filing. Again, firm bracing of the ring clamp against the bench is necessary. (See Fig. 14.)

At times several edges, too large for the ring clamp, must be filed simul-

taneously. Or again, a rather large single edge must be filed with great precision in preparation for a soldering join. Here the vise—with suitably protected jaws—becomes necessary. With work held in a vise, both hands are free to accurately guide the file over edges or surfaces. (See Fig. 15.)

When to use a large hand file or a smaller needle file is a question which only circumstances can answer. In general, it is more efficient to use a large file for large outside edges, whether straight or curved. A needle file may be used to further refine the work of the large file after the bulk of the filing is done.

The decision to bevel (angle or round off) an edge is best determined by the design. One design (perhaps one in which sharp crispness and angularity are emphasized) might have all edges at right angles to the flat surface. Another, perhaps more curvilinear, design might profit more by a softening of the edges. It is always amazing how great the illusion of three dimensionality can be when edges are beveled in a variety of ways. By beveling an edge one creates a greater variety of reflections in the polished final work which might enhance its richness. In addition, the sculptured quality beveling lends to a design removes it quite effectively from the feeling of stamped-out mass-produced jewelry. It affords the craftsman one more element of critical choice when deciding where and how far to alter an edge.

As a matter of craftsmanship it is well to soften the edges of the reverse side of a piece of jewelry. In this way it feels good to the hand and reminds one that good craftsmanship does not end with only the visible surfaces.

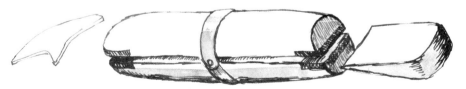

Fig. 14

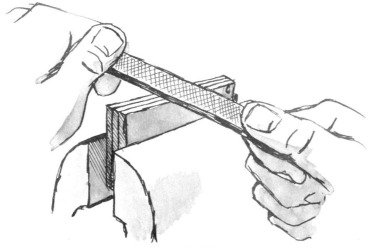

Fig. 15

Some of the needle files, in having a specific shape, have a function that should be exploited. As an example, the barrette file is smooth on two top surfaces and on its edges. This enables it to file close to an angle without cutting into the angle and causing additional work.

The round or rattail file may be used for enlarging drilled holes or in altering the shape of a drilled hole for decorative purposes.

The knife file is ideal for refining deep notches and angles whereas the triangular file may best be used for filing sharp notches and lines across a surface.

One of the pleasures of jewelry making is that each process is usually slow enough so that one may plan the next step while occupied with the first. This habit pattern develops a time- and energy-saving technique which soon comes with experience.

Since files in jewelry making are often used on a variety of materials—metal, wood, ivory, bone, etc.—they often become clogged with dust or metal particles. A file brush, preferably one with fiber bristles on one side and steel bristles on the other, is used to brush out most particles, and a narrow pointed scribe or pin may be used to pry out soft metal particles. It helps to dust the file with chalk since this does not prevent good cutting but leaves little space for other particles to become imbedded.

If a file is used on lead or lead solder, it should either be cleaned meticulously after use or set aside for this purpose in the future. A particle of lead which comes in contact with a high-temperature soldering process on silver or other metals will eat into that metal in a way that is impossible to repair without great effort. Keep lead away from silver soldering operations!

Scraping

In this process the metal is actually carved away. A scraper may be used, if one has strength and control, to cut angles or bevels on edges or to form depressions on flat surfaces. Most often this tool is used to remove excess and unwanted solder. With care the solder lump may be shaved away until the clean metal is exposed. Scraping usually leaves a surface that must be further refined by filing, stoning, or sanding, but it is quite a bit faster than the latter techniques used alone.

The scraper should be kept very sharp by honing it on a fine oilstone and the tip should be plunged into a large cork when not in use.

The scraping technique demands that the cutting edge be parallel with the surface to be scraped, for a higher angle would cause a "chatter" which forms ridges and roughness. (See Fig. 16.)

Stoning

In places where a file or a scraper would not fit easily, or where a design is too delicate for their use, a scotch stone becomes practical.

Scotch stones come in a variety of lengths and dimensions. For general use a stone ¼″ square and 5″ long is ideal. To enable stoning in narrow, restricted areas, the stone may be filed or ground to the necessary shape and kept to that shape by further filing or grinding.

Stoning should be done under flowing water or in a pan of water. If done while the stone and the metal are dry, the metal particles soon clog the stone. It is best to stone an area larger than the spot where solder or roughness must be removed. Don't replace a lump of solder with an equally unsightly hollow! Since the stone is quite smooth and also rather soft, considerable pressure is needed for effective stoning. A sharp jet of water or a little brushing will remove loose stone and metal particles after stoning. Check the stoned area often during the process by wiping away the residue.

Burnishing

A burnisher is a highly polished, hard steel tool with a handle. The blade is usually a pointed ellipse in section and may be curved or straight. A blade length of 3″ to 4″ is ideal for jewelry purposes and of the two shapes the straight burnisher is the more easily controlled. Burnishers should be stored in a wrapping of chamois and kept highly polished at all times.

The burnisher has two basic uses in jewelry making. The first is to smooth and give a high polish to beveled edges where other polishing techniques would not be practical. The second use is for removal of deep scratches and pits on surfaces. In burnishing out a scratch al-

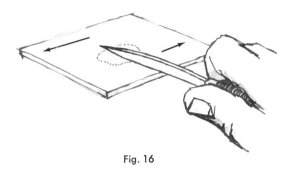

Fig. 16

ways rub the blade in the direction of the scratch. If rubbed across the scratch a dip results which might be worse than the original blemish. (See Fig. 17.)

Considerable controlled pressure is necessary in burnishing since the surface of the metal is actually moved and compressed to fill in scratches and pits.

Burnish over a large enough area to avoid forming a depression. A drop of light oil helps in moving the burnisher. Coarse emery may be used to remove burnishing marks.

A third use for the burnisher is to press a bezel around a gem stone. This will be described in detail in Chapter 4.

A word of caution: the high polish and considerable pressure on the burnisher can cause it to slip. This often creates a deep scratch, much harder to remove. Use short, controlled rubbing strokes and use the free hand as a support and guide when possible.

FINISHING

This is the most important technique mentioned so far since upon its effectiveness depends much of the final quality of the piece.

There are many forms of abrasive cloth and paper, but one or two types usually suffice. A fine and a fairly coarse *emery* paper have proved to be effective and economical.

In one brand, Behr-Manning, sizes Nos. 1 and 3/0 fulfill most needs. The No. 1 emery, being sharp and quite coarse, is used for the removal of file marks, scraper marks, scotch-stone marks, surface scratches and pits (if not too large), and minor beveling. If carefully handled it may be used to give an over-all matte finish to the completed work.

The No. 3/0 emery paper is used primarily to refine the surface left by No. 1 emery paper and to clean surfaces of oxides, oil films, fingerprints, etc., in preparation for soldering.

Both types of emery paper should be used in small amounts. A piece 2″ square, folded in half so that the rough surfaces are on the outside, is more efficient than a large handful. In folding the paper one prevents having it slip between fingers and metal during the pressure of sanding. Here again considerable pressure should be used for effective sanding.

Smaller pieces may be folded, rolled, or wound around an appropriate needle file so that narrow and difficult areas may be reached.

To prevent deep parallel scratches it is well to rub in several directions during sanding.

Do not discard worn pieces of emery too quickly! They might be useful where a finer sanding is needed.

CUTTING TECHNIQUES (BOBBING)

Materials

    Hand buffing sticks
    Cutting wheels
    Electric motor
    Cutting compounds: Pumice, powdered emery, Lea compound "C," tripoli

The craftsman need not use an electric polishing or cutting machine to bring metal work to a desired finish. For centuries before the use of electricity rubbing with coarse or fine powders by hand or with leather alone achieved very highly reflective and rich surfaces.

The cutting or bobbing operation consists of removing the marks of previous tool use (such as filing, burnishing, and sanding) by hard rubbing with a sharp but uniformly small-particled substance. This may be done by moistening a material like pumice to make a paste and using thumb and fingers for rubbing to a smooth finish. A faster, more efficient method is to use a wooden hand buffing stick, half of which is covered with felt or leather. These sticks may be purchased or a strip of the proper material may be glued to a suitable flat stick. In making the buff, one may design it for a specific purpose—i.e., thin and narrow for tight angles, round and tapered for small concave surfaces, sculptured to fit a special form.

Since water-moistened pumice would fall away from such a tool before much cutting could be done, a substance called tripoli may be used. Tripoli consists of a fine siliceous ooze in powder form which has been mixed with tallow or wax and pressed into blocks or bars. The tallow adheres the abrasive to the rubbing surface.

In hand bobbing considerable pressure must be used and one should avoid moving in one direction too long, for an unwanted groove may result.

Emery, an impure form of corundum,

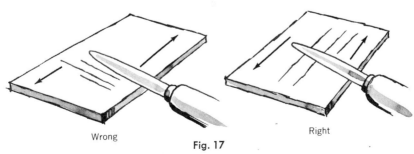

Wrong   Right

Fig. 17

the mineral of which sapphires and rubies are made, comes in the same forms as tripoli and may be used in the same manner. This is one of the oldest abrasive materials used by man.

In machine bobbing, the use of wheels of various materials attached to the spindle of an electric motor may save in time and effort, but an inexperienced worker might also bob away more metal than planned.

There are a great variety of wheel types available. Some do only one type of work well and others might do several.

Wood—Used for cutting primarily with emery or pumice as the abrasive. Special shapes may be formed to accommodate a specific need.

Leather—These wheels, while expensive, outlast cloth or felt wheels and hold the cutting compound better. The best wheels are of buffalo or walrus hide.

Felt—Available in a number of hardnesses and may be used with most compounds. Felt wheels are used most often on angular edges and small planes. It is quite easy to cut unfortunate grooves in metal by improper positioning while cutting!

Cloth—These wheels come in a variety of shapes, materials, and diameters. Some are nothing more than a number of disks fastened by a metal grommet in the center; others may be stitched and glued to give firmness to the shape. They may be made of cotton, muslin, or wool. Abrasives may be used with cloth wheels either dry (as tripoli) or moist (as in a pumice paste). By varying diameters and motor speeds a variety of bobbing actions may be achieved.

Brushes—Also available in a variety of materials and shapes. The bristles may be of fiber (Tampico brushes), pig bristle, or a plastic such as nylon. A Tampico brush used with activated pumice at slow motor speeds produces a faster cutting action than any other type of wheel.

Wheel mandrels—Mandrels are used to bob or polish the interiors of rings, bracelets, etc. They may be made of hard felt or of wood with a slot to hold a covering of emery paper or of emery stone. Emery paper of various grits is also constructed as a hollow cone to fit over a cone-shaped wood spindle and adhered by centrifugal force.

After constant use with tripoli and other cake abrasives, the surface of a bobbing wheel might become so matted

Fig. 18

that it not only cuts inefficiently but also might cut unevenly. Excess abrasive may be removed by holding a coarse file or hacksaw blade against the rotating wheel. Or the wheels may be washed in a solution of 1 quart (qt.) of hot water to which 1 tablespoon (tbsp.) of ammonia and a little salt have been added. The addition of a liquid detergent helps dissolve the binding medium, and the wheel may be spun dry after washing.

Effective cutting, buffing, or polishing depends as much on motor speed (rpm) as it does on the type of wheel or abrasive used. The larger the wheel diameter, the greater will be the speed at its working edge.

The surface speed, the rate of travel of the wheel surface past a given point, can be calculated by the following formula:

$$\text{Surface speed} = \frac{\text{Circumference of wheel in inches} \times \text{rpm}}{12}$$

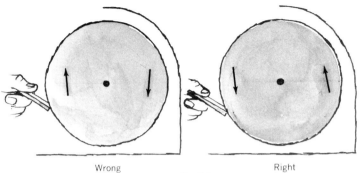

Wrong   Right

Fig. 19

The following are recommended motor speeds for a variety of wheels:

For leather or felt .... 2000 to 2500 rpm
For muslin or flannel .... 3000 to 4000 rpm
For bristle .... 1250 to 1750 rpm

An ideal motor arrangement consists of having the wheel area housed in an open-fronted hood with some means of vacuuming dust particles into a container. Lacking this, a simple hood for protection may be made of galvanized sheet metal and a pan with 1″ of water with a little detergent placed under the wheel. This collects a great deal of the waste. (See Fig. 18.)

The steps to safe and effective cutting are simple but important:

Step 1. Place the wheel on the revolving spindle for quick centering.

Step 2. Apply abrasive to the revolving wheel.

Step 3. Hold the work securely in both hands if possible. If the wheel catches a projection on its down swing, the work may be torn from the hand, causing damage to both the user and the work. The backdrop or hood is necessary since a piece of metal might be spun away from the wheel with great force. Always work with the wheel rotating forward and down. Apply the metal piece to a point below the horizontal median of the wheel. At this point the wheel rotates *away* from the work which should be angled to arrange this. Press the work against the wheel enough to notice and feel the cutting action. Light pressure is ineffective.

Step 4. Move the work constantly! If it is allowed to remain in one position too long, excessive cutting might remove detail and change the shape. Cross-polish for a uniform surface.

Step 5. Stop occasionally to reapply abrasive and to check progress. A piece of cleansing tissue may be used to wipe away abrasive waste. If the piece heats up through friction, it may be dipped into a pan of water occasionally. Small pieces too difficult to

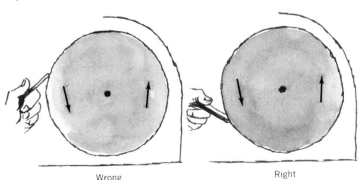

Wrong   Right

Fig. 20

hold by hand and which would heat up too quickly may be held in a ring clamp. Chain or other articulated forms must be firmly wrapped around or tacked to a board to prevent being fouled in the wheel.

Step 6. Clean all traces of bobbing compounds from work. If left in cracks or seams they may cause scratches during the final polishing process.

Trumming

This is an ancient method for bobbing and polishing delicate or intricate areas too small for hand or machine work. A small cord, preferably of nylon for long wear, is held by one end in a vise. The other end is held in the hand, pulled tight, and rubbed with an abrasive. The cord is then placed through or into the opening of the design and again stretched. Cutting or polishing is achieved by moving the work back and forth along the string with pressure.

STAMPING

After all parts are sawed out, filed, sanded, and bobbed, but *before* soldering starts, metals such as silver, gold, and platinum may be stamped with marks identifying quality, maker (Hallmark), or the word "Handwrought." Stamping later in the process of working the piece might be difficult or impossible without marring the design.

Most craftsmen today purchase such stamps as "Sterling," "14K," "18K," "Platinum," and "Handwrought." These come in a number of sizes and all but "Handwrought" must be displayed on work claiming to use that metal if it is to be sold. There are strict domestic and international laws governing this practice.

A hallmark may be made of soft tool steel by engraving, carving, or filing the design and then case-hardening the

stamp, or a stamp may be purchased made up to the design of the craftsman.

Number stamps and letter stamps are also available in various sizes.

The work to be stamped is placed on a *polished steel* surface and one or more blows with a hammer used to indent the end of the stamp into the softer metal. Care must be exercised so that the stamp does not move between hammer blows.

The slight blemish on the reverse of the point stamped may be removed by sanding or stoning.

Be sure to place the mark where future soldering of shapes or findings will not cover it!

SOLDERING STERLING SILVER AND OTHER NONFERROUS METALS

There are two basic soldering techniques to be used with sterling silver. Soldering done at temperatures above 1000° F is called "hard soldering" while soldering done at temperatures below 1000° F is called "soft soldering." The materials, steps, and even principles are totally different and cannot be interchanged.

Materials
    Solder
    Flux
    Flux brush
    Charcoal block, magnesium block, pyrofax or asbestos coil
    Tweezers: pointed and locking
    Pointer
    Iron binding wire
    Torch: gas + air, acetylene, propane, or mouth blowpipe
    Emery paper
    Pumice
    Brush

There are five sequences to accurate soldering. Each is as important as the next and performing any of them carelessly can result in failure. The steps in sequence are:

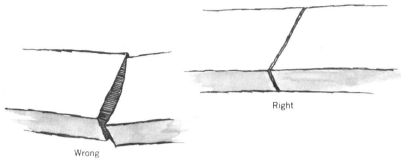

Right

Wrong

Fig. 21

Step 1. Fitting = Making a tight even join between surfaces to be soldered.

Step 2. Cleaning = Removing surface films of grease, oil, or oxide.

Step 3. Fluxing = Applying liquid or paste flux to all areas in sufficient amounts to prevent oxidation during heating.

Step 4. Solder placement = Placing the correct solder in the right places and in the right amounts.

Step 5. Heating = Using torches or blowpipes in a manner which quickly and safely causes solder to melt and flow.

Step 1. Fitting

Surfaces and edges to be soldered together must fit closcly along their entire length and breadth. Since flowing solder is, among other things, attracted by capillary action to a junction of surfaces but will *not* bridge gaps or irregularities, the close fitting of a join cannot be neglected.

In the soldering of two flat surfaces, both must be free of warping or dents. Large dents must be removed by careful hammering with a *wood* or *plastic* hammer over a smooth flat surface. Smaller dents may be removed by filing or stoning.

Warped metal may be flattened in the same manner or, if thin enough, may be corrected by hand. Sight along several edges to check final flatness.

Wire often develops kinks in shaping which are best removed by hand through bending and counterbending. Remember, solder flows along a join easily but cannot bridge sections out of contact with each other.

Joining ends in a butt join is simple if the ends meet tightly and evenly over their entire length. A butt join may be strengthened if both ends are beveled in order to increase the surface to be soldered.

If a warp in sheet metal or a kink in wire is very slight, the softening effect of heating might cause the piece to drop into position. This may be aided by pressure with a pointer or tweezers just as the solder begins to flow. Do not remove heat while doing this since the solder "freezes" almost immediately and will not flow between the new contacts.

It is far better to make an initial

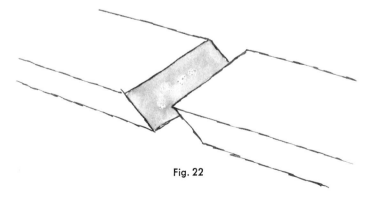

Fig. 22

proper fitting than to depend on the foregoing. If done unskillfully, press fitting of heated metal may shatter wire or sheet or the parts may move completely out of position.

Step 2. Cleaning

Solder will not flow over or onto a dirty surface. All traces of fingerprints, cutting or buffing compounds, and oxidation *must* be removed.

Chemical cleaning is most effective but is also time-consuming. In this process oils and greases are burned off by heating the metal to a dull red. Oxides are then removed by acid pickling and the work then washed thoroughly. If there are small hollow spaces where acid pickle collects, the acid must be neutralized by boiling the work in a solution of 1 tbsp. baking soda to 1 cup water. After rinsing, work may be kept under water to prevent oxidation and should be handled with tweezers to prevent new fingerprints.

A faster, and almost as effective, cleaning process involves brushing the work with moist pumice and a bristle brush. Work may also be cleaned by sanding it with a *clean* piece of emery paper. For most purposes the last is adequate for removing all dirt. A clean surface is bright and allows flux to be spread evenly without forming drops and pools.

Sanding wire after bending it into shape is difficult, so bending must be done with clean fingers or with pliers. If wire is to be soldered to a flat surface, it might be possible to sand its contacting edges clean on a flat sheet of emery paper.

Step 3. Fluxes and Fluxing

Hard solder flux may be boric acid, borax, or a prepared liquid or paste combining several fluxing ingredients. A lump of borax may be rubbed to a paste by adding a little water to a slate dish designed for this purpose. This is the flux-

ing technique used by some of the finest jewelers of the past. Today it is more efficient to use one of the several patented mixtures since they have greater fluxing and cleaning action. Some fluxes may also be used as temperature indicators.

It is important that *all* surfaces to be soldered are completely covered with flux. Each piece of solder should also be fluxed. This may be done by putting the solder fragments into place with a flux-moistened brush.

To prevent excessive cuprous and cupric oxides from forming on sterling silver, it is well to flux all surfaces except the bottom of the work. Since this is in contact with charcoal or asbestos the oxides do not form so easily.

The action of flux is to prevent oxides from forming when metal is heated. Borax, or fluxes containing borax, become fluid and glasslike around 1200° F and thus prevent oxides or dissolve oxides as they form.

Prolonged high heat breaks down the protecting qualities of fluxes to the point that they will no longer absorb oxides and may even pull away from the surface entirely. This is the reason that heating must be done quickly and only enough to cause solder flow.

Step 4. Solder and Solder Placement

Hard solder is an alloy of the metal on which it will be used with small amounts of other metals added. Thus silver solder is an alloy of silver, copper, zinc, and cadmium in varying amounts. By combining metals a new metal is formed that usually has a lower melting point than the metals used to form the alloy.

If more than one soldering is needed to complete a piece, different solders may be used with melting points that are successively lower. Handy and Harman supplies this list of melting points for its silver solders:

"Easy" solder (for most simple soldering) 1325° F
"Medium" solder (for first or second
solderings, or multiple soldering op-
erations) . . . . 1390° F
"Hard" solder (for first or second solder-
ings, or multiple soldering operations) 1425° F
"IT" solder (only for joins requiring
great strength) . . . . 1460° F

Remember that sterling silver begins to break down at 1500° F and becomes liquid at 1640° F. Fine silver is liquid at 1761° F. Hard silver solder comes in sheets, wires, sticks, strips, or powder forms. For jewelry purposes sheets or strips are the most practical.

The sheet or strip is cut into small pieces or *paillons*. Since it is usually better to use several small, easily heated pieces rather than one large piece at a

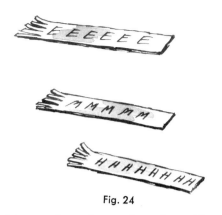

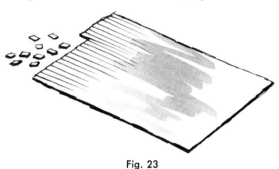

Fig. 23

given point, these *paillons* should be about ⅟₃₂″ square. Even smaller pieces must be used when soldering fine wire or small shot. Scissors or jeweler's shears are used to "fringe" the edge of a sheet of solder for a depth of about ½″. The cuts are ⅟₃₂″ apart and the pieces are formed by making a right-angle cut ⅟₃₂″ wide across the fringe. (See Fig. 23.)

Place the index finger along the side of the scissors to prevent the solder from flying around during cutting.

To prevent mixing solder pieces of different melting points, each solder should have a separate, labeled container. A low, wide-mouthed cosmetics jar is ideal. In addition, each sheet or strip should be scratched with the initial of the solder type to prevent confusion. The *paillons* should always be much

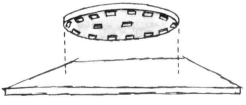

Fig. 24

thinner than the work to be soldered. Usually 28 to 30 B & S gauge is thin enough, but it may be rolled or hammered thinner if necessary.

Solder should be placed only where, in flowing, it will make a good join. Use only enough to do the job. Too much solder causes difficulty in melting, as well as blemishes. If excess solder flows off a join onto a visible surface, it may prevent future evenness of coloring and must be removed by stoning. Excessive solder while soldering wire might prevent some of the solder from flowing.

SOLDERING FLAT SURFACES

After determining which piece will be on top, and after thoroughly cleaning and fluxing all surfaces, place the solder on the *reverse* of the top piece; this insures that all solder will be covered correctly. The solder should be positioned ⅛″ apart around the outside edge of the top piece with an additional three or four pieces in the middle if the surface is large. (See Fig. 25.)

Before lifting the top piece with tweezers, allow the flux to dry for a moment. This will prevent having the solder fall off or move while upside

Fig. 25

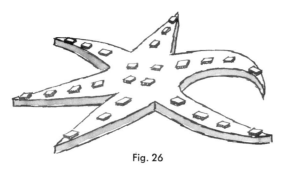

Fig. 26

down. The *paillons* should be put into place by picking them up with a flux-moistened brush tip. Choose a small, pointed brush and let it be inexpensive —one accidental application of solder to a hot piece of metal and the bristles are gone!

Where the top piece of metal has long projections, make sure that solder is placed at every tip. If these projections are narrow, do not place *paillons* all around the edges but only down the center of the projection. (See Fig. 26.)

On flat to flat or slightly curved surfaces soldered horizontally, it is not necessary to use binding wire to hold work in position. Gravity will keep even small pieces in position and the wire binding will become too loose to do much good once the solder flows.

The two surfaces should be separated only by *paillons*. Be very sure that no foreign matter is between surfaces or the work will not come together when solder flows! (See Fig. 27.)

It is possible to solder two or more layers together in one operation, but when they are stacked too high it is

difficult to heat all of the pieces evenly. The top surfaces may be too close to melting while the bottom piece is many degrees cooler.

The work should now be placed on a charcoal or asbestos block and again checked for alignment. All displaced solder should be put back into place or removed. Materials such as charcoal or asbestos retain and reflect heat, thus speeding the soldering operation. In addition, charcoal tends to prevent oxidation where metal touches it. Such materials as brick, fire brick, sheet rock, and stone are unsuitable since they dissipate so much heat that quick, even heating of metal is impossible.

Step 5. Heating

Before describing the correct application of heat, it might be well to explain torch and flame qualities and how they should be used.

Gas-Air Torches

In established workshops and in many schools, torches are fed by a combination of gas and compressed air. The gases may be manufactured illuminating gas, natural gas, or propane. Special tips may be needed in each case since a tip for illuminating gas would not form an adequate flame when used with natural gas. Air pressure may be supplied by a foot bellows or by a motor-driven compressor. The torches, then, have two adjustable inlets for gas and air, and by

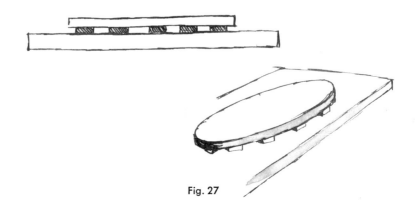

Fig. 27

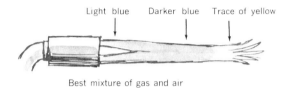

Best mixture of gas and air

Fig. 28

changing their proportions a number of flame sizes and types may be formed. Most torches also have a variety of tips available which form flames ranging from large soft flames to small pointed flames.

With a gas-air torch the most effective flame for soldering pieces 1″ in diameter is about 5″ long, is soft at the tip, and might have a *little* yellow at the end of an otherwise blue flame. Too much yellow—a reducing flame—is too soft for concentration of heat and might deposit unburned soot particles on work which could prevent solder flow. A too-yellow flame also lacks the heat necessary for quick soldering.

A flame which is a hard, pointed blue —an oxidizing flame—is even more dangerous. Far from melting solder quickly, it causes heavy deposits of cuprous and cupric oxides on sterling silver and other copper-containing metals as well as on the solder itself. This quickly prevents all solder flow. Being small-pointed, this flame also may overheat small areas instead of uniformly heating the whole unit. (See Fig. 28.)

When soldering small units such as chain links or filigree, the flame size should be smaller while using the correct gas-air mixture. Generally, the last inch of the flame should touch the metal surfaces.

ACETYLENE TORCHES

Self-contained pressure tanks of acetylene are available in sets consisting of a tank, hose, pressure gauge and valve, and an adjustable torchhead.

Acetylene gas has the advantage of forming a clean, quick-heating flame which may be used large or very small for a variety of soldering purposes. Being self-contained, it may be moved from place to place for use in several shop areas. The initial cost is high since the tank is bought outright or leased, but with normal use a full tank of gas lasts for quite a long time and may be exchanged for a refill at low cost.

Since the gas mixes with air at the torch tip without adjustment being possible, only the flame length may be varied, though tips of different sizes and shapes are available.

A No. 1 tip in the Prest-O-lite unit is good for jewelry work.

PROPANE CANISTER TORCHES

Small self-contained and disposable cans of pressurized propane gas are available with a reusable torch attachment. These are adequate for small jewelry work, but for prolonged heating or work on large pieces they do not last long enough to be economical.

The flame is not very adjustable, but it is clean and hot. Propane torches are worthwhile where jewelry making is of limited or occasional interest since most are sold for less than 10 dollars.

Bunsen Burners or Alcohol Burners

For many years craftsmen have used a simple brass blowpipe through which air is blown, deflecting and concentrating the flame of a gas Bunsen burner or alcohol flame. By these means a small, hot, and carefully controlled flame may be focused closely when delicate work is being soldered. An alcohol burner and blowpipe may well be used for soldering small gold or silver areas even when other torches are available. The flame is too small, however, to work well upon work over 2″ in diameter.

In areas where manufactured illuminating gas is still in use for cooking and heating, a torch similar to the above is both inexpensive and effective. It consists simply of a length of hose, one end attached to a gas jet (such as for the burner of a kitchen stove) and the other to a mouth blowpipe. The gas enters the pipe after the air through an interior pipe and mixes at the tip. With such a torch it is simple to adjust the size, the heat, and the duration of the flame. With practice in breathing in through the nose and using the cheeks

as a pressure chamber, it is possible to maintain a very even air supply and a steady flame (see Fig. 29). For many years the author used such a torch for all soldering and regrets that he now lives where it cannot be used, since it will not work well with natural gas.

Where possible, a gas-mouth blowpipe may be used at the kitchen stove, a sheet of transite providing a safe soldering area over the burner.

There are various small alcohol and chemical tablet torches available which do not supply enough of the right kind of heat to be practical. Electric soldering guns or irons are used only for soft (lead) soldering and even for this purpose cannot be used delicately in jewelry making.

Electric arc soldering machines are too limited in the type of work they can do, as well as being expensive, to warrant purchase by the jeweler-craftsman. For most conditions, the author prefers the self-contained acetylene torch and considers this a worth-while investment.

Correct heat application, with any of the above units, is obtained as follows:

The flame should first be rotated around the *outside* edge of the fluxed work without touching the metal itself. Metal heated too quickly causes even

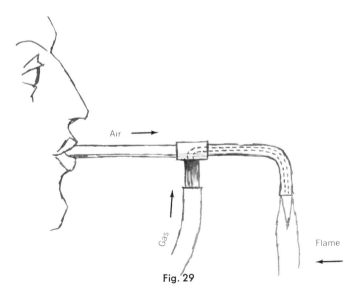

Fig. 29

dry flux to boil or foam, causing solder or even the metal pieces to move. When this occurs, both the pieces and the solder must be moved back into position quickly. Always have a pair of pointed tweezers or a pointed steel rod in hand for this purpose. If, upon cooling, the displaced pieces stick, the whole unit should be heated again slightly to remelt the flux, since hardened flux may hold solder and work very securely.

The best indication that foaming has stopped is the formation of a white crust on fluxed surfaces. Once the crust is formed the work may be heated directly.

The *entire* metal area *must* be brought to solder flow temperature as a unit. If heat is concentrated at one point, the surrounding metal will dissipate it away, prolonging or even preventing solder flow. Move the torch slowly back and forth or around the entire area until solder begins to melt.

There are several ways in which this may be determined. Some fluxes, such as Handy and Harman Paste Flux, become clear and liquid at 1100° F. When this happens one knows that only a few hundred degrees more of heat will cause the solder to melt.

With experience the color of the metal itself becomes an accurate guide. Many craftsmen solder in subdued light the better to see the medium dark red glow which indicates that solder should be flowing. Often it is possible to see the solder melt and flow, as in soldering wire, but when *paillons* are hidden between two pieces it is best to watch for (1) a drop of the top piece as solder changes from solid to liquid and (2) after a little additional heat a bright seam of molten solder showing at an edge.

If this seam is not noticed and if, through inexperience, the color of metal gives no indication that solder should be flowing, solder flow may often be incomplete. This often happens when soldering overlaid work. The top piece has become hot enough to attract the solder beneath it, causing it to flow to this surface. The top piece has dropped so it seems to be together—only to fall apart when pickled, or even later. To prevent this, always heat for a few seconds longer after seeing the top piece drop. This may very well bring some of the molten solder to the edge for a sure indication.

A number of things can cause soldering failures:

1. *Incomplete solder melting.* Cure: After careful pickling, all work is refluxed, solder added where necessary (if any has fallen off), and heat reapplied.

2. *Solder balling and refusing to melt.* Cause: Metal surfaces may be dirty with grease, oil, or oxides. Solder may be dirty. Excessive preheating may cause oxides or cause flux to be supercharged with oxides, thus no longer absorbing or preventing oxidation. Too much heat concentration may cause metal to cool in unheated areas. An incorrect flame (too much or too little air) may cause soot or oxides. Heating too slowly even if the correct temperature is finally reached (flux can prevent and absorb only so much oxidation). Prolonged high heat causes trouble. Incomplete solder contact; since solder flows to both the *hottest* area as well as to a junction of surfaces, it should be in contact with junctions and generally heat should be directed toward them.

3. *Foreign matter.* Through accident or carelessness particles of charcoal, asbestos, or even metal filings might remain between two surfaces, preventing them from coming together when solder flows. Always check for dirt and use *only* what you know to be solder.

4. *Accidental use of wrong solder.* If solders of different melting points are not kept in separate labeled con-

tainers, it is easy to make the mistake of using a high melting solder where only a medium or low solder can be used. Delicate work can be melted completely before the mistake is suspected!

5. *Careless use of whiting or yellow ocher.* Both of these substances (their proper use to be described later), when accidentally mixed with flux, can completely prevent solder flow. Keep fluxed areas separate from those painted with these materials!

6. *Poor fitting of parts.* This causes most difficulties since the solder cannot bridge gaps between surfaces or flow along wire when it is not completely in contact with its support. Though it is sometimes possible to press a warped sheet or wire into place *while applying heat,* great damage may result if work becomes too hot while doing so. Remember, most metals become brittle at high temperatures and can be shattered by forcing them. It is best to be sure of correct fitting from the outset.

7. *Moving of parts before or after solder flow.* As mentioned, if parts move before solder has begun to flow, they may be pushed back into position. After doing so be sure that all solder is also in position since a blemish caused by misplaced solder is very visible and difficult to remove.

If work moves just as solder melts (this often happens due to capillary attraction when soldering very small pieces close to each other), it is best to use tweezers or a pointer to move them while the solder is still fluid. Do not remove heat since this would cause the solder to solidify immediately, making remelting quite difficult. If a moved piece refuses to be repositioned, do not heat for longer than one try at correction. Continued high heat causes a stronger weld as well as burning away some of the zinc or cadmium in the solder. This causes it to melt at an even higher temperature as well as making the join porous and weak. It is best to pickle work carefully, to wash off pickle carefully (making sure that no acid remains between the surfaces to be soldered), and again to reflux thoroughly. If heat is now applied correctly the pieces *may* be moved into position, but usually it is more difficult to remelt solder which has once been fluid.

It is inevitable that, upon moving a piece to a new position, a solder scar, or "ghost," appears where solder has melted originally. This must be removed by abrasives or it will leave a roughness which is difficult to hide or color in the finishing process.

After solder has melted satisfactorily, the piece may be dropped into pickle while still hot or, if allowed to cool, may be boiled in pickle until clean. If dropped cold into cold pickle it will eventually be cleaned of oxides but this might take some time, depending on the strength of the solution.

Since it is possible to solder many times on the same piece, it is very important that oxides formed during one heating are all removed before the next heating. It is, of course, necessary to reflux everything carefully before each soldering! Be sure that no pumice or other abrasives work into areas to be soldered.

As a check list for successful soldering, remember to have:
1. Close fitting joins
2. Clean surfaces
3. Careful fluxing
4. The right amount, placement, and type of solder
5. Removal of foreign objects
6. Careful but quick preheating
7. Even heat of the correct sort over the entire metal area

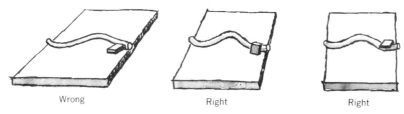

Wrong   Right   Right

Fig. 30

8. Enough heat to cause complete solder flow
9. Correct pickling or cleaning *and* refluxing between solderings

In general, the rules of hard soldering apply equally to soldering flat sheet to flat sheet, curved metal to curved metal, edges to edges, wire to flat metal, wire to wire, grains to flat metal, or grains to wire.

There are, however, a few basic techniques in working with wire or grains (shot) which should be mentioned.

SOLDERING WIRE TO SHEET

The basic difference between this and other soldering lies in the position and the amount of solder. Since only a small portion of a section of wire contacts another surface, only a very small amount of solder is necessary for firm soldering. Too much solder, if all the pieces melt at all, causes a flooding of wire edges which will give a blunt heaviness to a design.

Keeping in mind that solder is attracted to a join, one should place the solder in contact with the wire more than with the sheet. (See Fig. 30.)

Having placed the *paillon against* or on top of the wire, there is greater contact with the wire and, when the correct temperature is reached, it will flow down the sides of the wire and for quite a distance along the join.

A flowing piece of solder may travel for an inch or more along an 18 gauge sterling silver wire. Consequently, it is best to use very little solder and only where it will do the most good. (See Fig. 31.)

Note that the solder in Fig. 31B is placed at a junction of the wire and an edge. Should the solder flow onto the sheet rather than along the wire, the lump may easily be stoned away. It would be more difficult to do so were the solder placed somewhere in the middle. The two pieces of solder in Fig. 31B are both on one side of the wire. By angling the flame from the opposite direction (arrows), the solder flows *to*

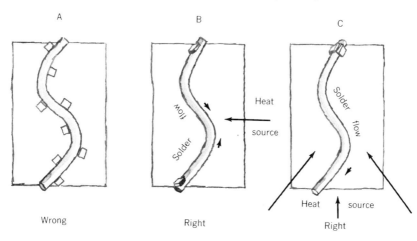

Wrong   Right   Right

Fig. 31

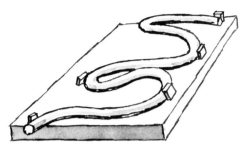

Fig. 32

the wire and the hottest area. This flame angle should only be reached just as solder begins to flow or else the work may be unevenly heated. It is possible to "draw" flowing solder along a join with the torch by the same principle (Fig. 31C).

Where a long length of wire is used, additional solder should be placed on outside curves—it is easier to remove mistakes—and on *top* of the wire. (See Fig. 32.)

It is perhaps easier to solder thin wire to sheet than thick wire. This is because the heat flow between wire and sheet is almost immediate when the wire is small. With thicker wire, 14 gauge and up, heating should be as for overlay soldering. In all cases, if the wire is in close contact for its total length, there is no chance of melting the wire before the sheet itself becomes overheated. Any point where the wire *leaves* the sheet might be the point where wire melts and breaks apart.

Where wire leaves the support and contact of sheet, as shown in Fig. 33, melting of wire is very possible. Adjust the flame size and the heating area accordingly and watch the color of the wire at all times. By alternately applying and removing the flame in an even manner, the total mass may be heated before projections become overheated. (See Fig. 33.)

An alternative is to place the wire on the charcoal, which would tend to distribute heat more equally. The difficulty with this approach lies in the placement of solder. It would have to be sandwiched between the wire and the sheet after having been applied to the wire first. (See Fig. 34.)

SOLDERING WIRE TO WIRE

This technique is similar to the above in that very little solder is used in order to prevent clumsiness and heaviness.

A smaller flame is necessary since a large flame would tend to overheat areas of wire not under close observation. In most cases it is best to use a high melting solder such as Handy and Harman "Hard," since this does not become quite so fluid as lower melting solders. Because of this quality, it makes a stronger and better-looking butt or angle join by forming a fillet. (See Fig. 35.)

The best possible fitting of wires is important for strength and appearance. Wire which has merely been nipped off with side or end cutters presents a crude ending which might destroy the linear continuity of wire. It also presents a poor surface for joining an end to another unit. (See Fig. 36.)

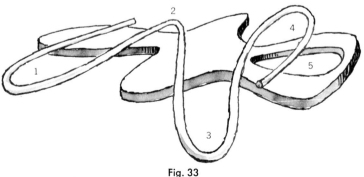

Fig. 33

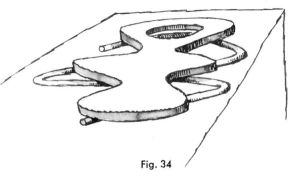

Fig. 34

In Fig. 36C, the end of one wire has been filed parallel to the join as well as concave to fit the curve of the other wire. This may be done with the round needle file. On an angled join, appearance is greatly improved by filing an angle to flow into that of the join. (See Fig. 37.)

Placement of solder for either of the foregoing should be economical. One piece should do the job or the join will be bulky, leaving excess solder to be filed away. (See Fig. 38.)

Allow heat direction to pull the flowing solder to the join. If the join does not meet, the solder will jump to one or the other wire—whichever is the hotter.

In soldering wires parallel with each other, if possible always place the solder at the ends only, in order to prevent "ghosts" along the length. (See Fig. 39.)

When soldering tight spirals of wire

to a flat surface, it is best to premelt the solder where the spiral should go first, then pickle, reflux, add the spiral, and reheat. In this way the solder will already be in firm contact with the base, instead of melting into and around the spiral only.

Open spirals need solder placement similar to that for straight or curved wire.

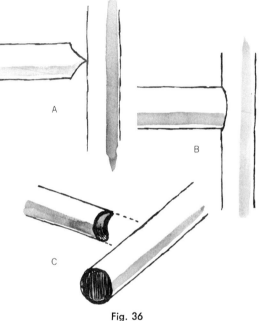

Fig. 36

SOLDERING SHOT OR GRAINS

If shot is to be soldered to a sheet surface, the solder may either be melted partially at the solder point, after which the fluxed shot is applied, or the solder may be applied with the shot as shown in Fig. 40.

If many shot are to be soldered in a row, it is best to engrave a guide line into which they may be placed. This should prevent the occasional moving of shot as the solder melts.

Where a cluster or textured surface of shot is needed, the area they are to occupy may first be coated with a film of melted solder. This should be done carefully since overheating of melted solder causes the alloy to change. This usually

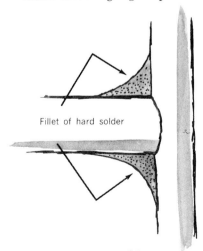

Fillet of hard solder

Fig. 35

43

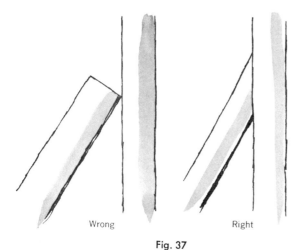

Wrong   Right

Fig. 37

prevents remelting. After the sheet, with the melted solder in position, has been pickled, rinsed, and refluxed, the fluxed shot may be put into position and soldering completed.

SOLDERING AT ANGLES

Perhaps the most difficult soldering operation—from the standpoint of easy solder flow—is to solder a vertical plane to a horizontal plane. The greatest difficulty lies in supporting the vertical well enough to withstand flame pressure and expansion. Here soft iron binding wire may be useful (be sure to remove it

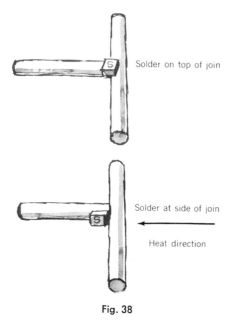

Solder on top of join

Solder at side of join

Heat direction

Fig. 38

44

completely before pickling!) for holding parts in place, but usually a good job of edge filing allows a piece to stand unsupported.

If possible, solder may be placed on the edge and between the vertical and the horizontal plane. As the solder melts, the vertical drops into place and may fall over unless supported for a moment with tweezers or a pointer. (See Fig. 41.)

A safer technique (which may, however, cause a few "ghosts") is to place the *paillons* along one side of the vertical, edging against it but lying flat on the horizontal. Placed this way, there is less tendency for the solder to flow up the side of the vertical rather than along the join. Use long thin *paillons* and avoid

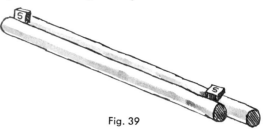

Fig. 39

heating the vertical plane too soon. (See Fig. 42.)

Similar principles govern soldering more acute angles, but they should be braced to prevent moving and falling during soldering. Pins, bent to shape, of nichrome wire (used in heating elements) may be pressed into the soldering block for this purpose.

When using liquid fluxes, always avoid an excess. This, even after drying, may cause boiling which could move delicate work. After an area has been thoroughly fluxed, the liquid excess may be drawn off with a blotter or cleansing tissue. Enough remains to do the job.

OVERHEATING STERLING SILVER

The possibility of overheating is always present during soldering. Often one's attention is fixed so strongly at one point that accidental heat concentration at another point only a few mil-

 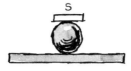 

Fig. 40

limeters away may cause the metal to melt unnoticed. This is especially true while soldering wire forms because of the small heat-dissipating area.

Silver that is being overheated goes through several steps, all of them serious:

1. The color changes from red to an incandescent red orange.

2. The surface becomes shiny as it begins to melt.

3. The form begins to lose its solidity, often warping, sagging, and melting at edges.

4. The edges draw in and eventually the entire piece collapses into a large circular mass of molten metal. Wire might burn apart and draw away from the break.

Once silver has reached Point 2 above, it has changed its structure radically and may no longer be bent or formed, being much too brittle. In addition, the surface has become granular and roughened. This may sometimes be repaired by careful filing, sanding, and burnishing, but, due to excessive cupric oxide "fire" deep within, it will always look different from areas not overheated.

Overheating also causes weakened joins due to the burning out of zinc and cadmium in the solder. Once gone, they leave a porous metal composed mostly of silver and copper oxides.

SOLDERING GOLD AND KARAT GOLD

The same heat units (torches, etc.) may be used for soldering gold as for fine and sterling silver. Since gold solders flow at a somewhat higher temperature than silver solders, a prepared flux should be used that functions well at such temperatures.

The principles of heating, solder size, placement, and pickling are also the same as for silver.

Only the gold solders and their actions introduce new factors. Solders for gold are, in most cases, karat golds of the same color as the basic metal but about 4 karats lower in purity. Thus, to solder 14K yellow gold, a 10K yellow gold of identical color is used as the solder.

Since gold solder never becomes quite so liquid as "Easy" silver solder, an even lower karat of gold must be used where a long flow, as for soldering wire to sheet, is required.

One of the advantages of gold solder over other types is in its slow flowing

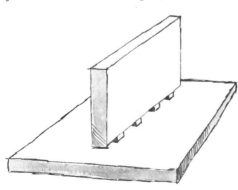

Fig. 41

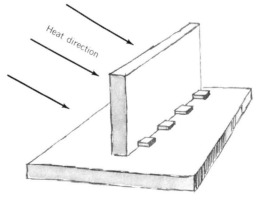

Fig. 42

tendencies. Occasional holes or gaps may be bridged by adding solder while heating. At the right moment heat may be withdrawn, leaving the bulk of solder in place. This leads to the temptation to overload joins. As in silver soldering, the strongest and best-looking join uses only enough solder to join all necessary surfaces.

Gold may be pickled in a 10% sulfuric acid pickle or in a special gold pickle of 8 parts water to 1 part nitric acid. This should be used in a pyrex or porcelain bowl.

Color is *not* a good indicator of heat when soldering gold, for gold shows little color until its melting point is reached and then it collapses quickly.

White, yellow, and green golds may be either air-cooled or quenched without affecting softness or hardness. Red golds tend to harden on slow cooling. They should be quenched while red hot. These factors are also important while *annealing* gold.

SOLDERING COPPER, BRASS, AND BRONZE

In the soldering of these metals, the rapid and heavy oxidation of copper presents considerable difficulty. If preheating is not done quickly and carefully, the fluxes will have exhausted their absorbing and protecting abilities and the solder will "freeze." As in silver and gold soldering, all surfaces must be free of oxides or solder will not melt and flow. Because of heavy oxidation, flux should be applied as thickly as is practical.

Heat should be supplied rapidly to quickly bring the metal up to the solder flow point.

Silver solders may be used for copper, brass, or bronze, but the difference in color may be disturbing. Solders somewhat closer in color are called *spelters*, and their use, *brazing*. Spelters are a copper-zinc alloy, 33% copper plus 66% zinc, for an easy-flowing but fairly strong join, but they cannot be overheated because the zinc tends to burn out, leaving roughness and pitting.

When soldering brass or bronze to silver, it is very easy to overheat slightly, thus causing the base metals to collapse. They quickly alloy themselves with the silver around them and melt into the silver, destroying shape and detail. When making this combination, use the lowest melting hard solder or spelter available and heat very carefully.

Aesthetically and practically, it is poor practice to use lead base *soft* solders on jewelry. At times it is, however, necessary to do so to prevent difficulties. In attaching some of the findings and also sometimes in combining brass with silver, soft solder is almost the only practical material to use.

ADDITIONAL SOLDERING INFORMATION

1. When soldering near an already soldered vertical join, the join may be protected with a paste of water and yellow ocher, an iron oxide. Powdered rouge, whiting, and clay may be used for the same purpose. The paste on a join "dirties" it and prevents an easy solder flow. Do not depend on this too heavily! An overheated join will collapse *under* the yellow ocher coating at high temperatures.

Always be very sure that flux and yellow ocher have not combined near the new join. One cancels out the other.

Yellow ocher is removed when work is pickled but it may be very stubborn if overheated.

2. Exact placement of shot may be achieved by using a small dapping punch to indent a shallow depression in which the shot may rest. Small shot moves about easily as the solder melts, due to capillary action.

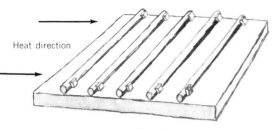

Heat direction

Fig. 43

3. Straight parallels of wire also have the tendency to jump together as solder flows. A slightly curved wire is, of course, very stable, but if a straight section is needed in a design four techniques may solve the problem of movement:

(a) Solder may be placed only on one side of each wire, the heat angled to draw the flowing solder only to its own wire. (See Fig. 43.)

(b) A clean shallow line may be incised with a burin or scorer and the wire placed into it. (See Fig. 44.)

(c) On vertical or near vertical surfaces a few *stitch* marks may be used to supply a brace for wire. A stitch is a sharp-angled cut made with a burin which leaves a prong or burr above the surface. These should be small, so that they will not show after soldering. (See Fig. 45.)

(d) Since a previously melted join does not flow so readily the second time, each wire may be soldered on individually (with pickling and cleaning between each soldering) and thus reduce the chance of wires moving on subsequent solderings.

4. Cupric oxide (the deeply penetrating purple or black oxide in sterling silver or copper-gold alloys) may be prevented by: (a) Dipping the grease-free, warmed work into a saturated solution of boric acid and alcohol. The work is again heated which ignites and burns off the alcohol, leaving the piece covered with a thin borax film. Handled carefully during fluxing, solder application, etc., it may be soldered without the formation of cupric oxide. (b) Heating the work to a light brown color and dipping it into a solution of 6 tbsp. boric acid + 6 tbsp. potassium fluoride to 1 pint (pt.) of water. After removing and drying, the work may be prepared for soldering as usual. Do not inhale the fumes during soldering since potassium fluoride might cause nausea.

When cupric oxides occur they may be removed in an acid bath. See "Removing Deep Cupric Oxides (Bright Dipping)" (page 49).

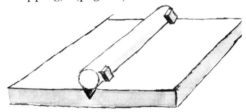

Fig. 44

5. Paste fluxes have qualities and disadvantages over liquid fluxes which should be taken into consideration. Paste fluxes prevent oxides better and longer during heating but they tend to become so viscous that solder pieces and parts themselves may float about during heating. During preheating they foam enough to dislodge and displace parts, which may not be noticed until they have melted in the wrong position. After solder-melting heat has been reached, a *borax glass* is formed on the metal which may be removed only in hot water or pickle.

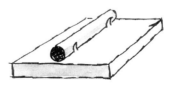

Fig. 45

Liquid fluxes—if used skillfully—work almost as well as paste fluxes, do not foam or bubble excessively, and are easily removed when work is dropped hot into the pickle bath. Their disadvantages—especially for the beginner—lie in the short fluxing action at high heats. Liquid fluxes exhaust their protective films quickly and prolonged heating may result in heavy oxidation of metal.

SOLDERING FOR FERROUS METALS

Though not often used in hand-wrought jewelry, iron or the many forms of steel may need soldering at times.

The soft solders (lead base) may be used on almost all ferrous metals, but they do not make a very strong join. Silver solders, because of rather high flow points, might cause too much annealing of the ferrous metal parts.

Brass wire, usually quite pure, or a special spelter for iron or steel, may be used as a solder with a borax base flux.

Work may not be pickled in nitric or sulfuric acid solutions because of the strong reaction on ferrous metals. Borax and other fluxing residue may be removed by hot water and oxides may be removed by abrasion.

PICKLING AND CLEANING

Several times during the soldering of a piece of jewelry, the excess oxides and used flux must be removed before soldering may again proceed. Though this may be done by filing, sanding, or wheel cutting, usually it is achieved by pickling.

Pickling

The pickle for gold, silver, copper, brass, bronze, etc., is composed of 1 part sulfuric acid added to 10 parts cold water. Remember to *add the acid to the water* while stirring with a glass rod, be-

cause water, when added to many acids, results in great chemical heat—even explosion.

Work may be dropped into pickle while still hot after soldering or it may be put into a solution while cold. When the pickle is then brought to boiling, the work will be clean of *cuprous oxide*. Cuprous oxide is the surface black or gray oxide. The cupric oxide, usually red or purple, penetrates deeper and is not removed by sulfuric pickles. See "Removing Deep Cupric Oxides (Bright Dipping)" (page 49).

Cold work, if left in cold pickle long enough, eventually will be cleaned, but solder joints may also suffer when zinc in solders is dissolved.

Pickle may be stored in a pyrex jar or a covered stoneware crock. A sheet lead vat, large enough for hollow ware pickling and supplied with a steel plate beneath the bottom, may be used in a bench. A bunsen burner beneath the steel plate supplies heat when hot pickle is needed. Small amounts of pickle are best heated in a copper pan with a handle. Continued use of sulfuric acid will eventually corrode a copper pan, so it is wise to rinse it in water after each use and also to dust it with sodium bicarbonate. It should be rinsed again before use since the sodium bicarbonate would neutralize the pickle.

Steel or iron of any sort should *never* come in contact with the pickle. Once iron has touched the pickle a galvanic process deposits copper on all work put into it. Use only copper, nickel, or brass tools to remove work from pickle. Work may be dropped in from a short distance when held in steel tweezers. Remember that pickles are very corrosive, so avoid splashing. Very hot pieces should not be dropped immediately into cold pickle since the shock of rapid cooling may cause the metal to shatter. Let everything cool to a dull red before immersion.

Rinsing

This must be done after each pickling to prevent corrosion of the metal when heat is again applied. If done carelessly, skin and clothing can be burned by acids. Flush all areas as well as possible and if the work includes pockets and hollows these must be neutralized by boiling in a solution of 1 heaping tbsp. sodium bicarbonate to 1 cup water. If this step is neglected, the pickle could continue to corrode for a long time.

If an overlay of two sheets is incompletely soldered, all acid *must* be washed out between them before resoldering, since the acid salts formed during reheating would prevent solder flow.

Usually, after the first or second pickling, the surface of sterling silver work becomes frosty white. This surface is a very thin layer of pure silver left after copper oxides have been pickled away; it may be preserved as it is or it may be carefully *burnished* to a high polish. Any abrasion or mechanical bobbing will, of course, quickly remove it.

Where a solid, dark patina of silver sulfide is desired (oxidation), the pure silver deposit must be removed, by scrubbing with a brush and fine activated pumice. This tends to cover the surface with many fine scratches, but if the pumice is fine enough this should not cause great polishing difficulties later. It is a waste of effort to highly polish work before soldering. If the surface is uniform and free of pits and scratches, polishing is simple after soldering and pickling have been completed.

A pickle for gold is made up of 1 part nitric acid to 8 parts water, and is used cold.

Removing Deep Cupric Oxides (Bright Dipping)

After *all* soldering has been completed, the cupric oxides, usually noticeable as a dull red to purple blush or spotting on sterling silver or 14K golds, should be removed. When done mechanically by wheel or hand abrasion, fine detail and crispness of design are often lost.

The oxides might be hidden by heating the metal to medium red and quenching in pickle repeatedly. This leaves a thin layer of pure metal on the surface. Great care must be exercised in buffing and polishing in order to avoid breaking through this layer to the oxides.

A uniform and quite durable finish using the above technique may be developed this way.

1. Polish and clean the work after all soldering is done.
2. Polish with rouge once more to bring out the cupric oxide scale. Clean thoroughly.
3. To develop the cupric oxide scale uniformly throughout the work, heat it to about 1200° F (dull red), and allow it to cool slowly in the air.
4. When cool, pickle the work in hot sulfuric pickle for 1 minute.
5. The pickle dissolves the surface oxides, leaving a film of fine silver or fine gold.
6. Burnish the matte surface of the piece to a soft luster, using a fine brass brush revolving at 800 rpm with soapy water as a lubricant. (Some craftsmen use stale beer for this purpose.)
7. Repeat the entire process of heating, pickling, and burnishing at least three times to develop a durable surface of fine silver or fine gold.

The best technique for the average jewelry maker is to dip the work into a *bright dip* consisting of 50% nitric acid added to 50% water and used cold. The work is carefully cleaned by pumice scrubbing and fastened to a length

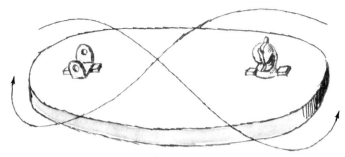

Fig. 46

of stainless steel wire. (Copper wire, though used in normal pickling, would soon be dissolved.) The work is dipped into the bright dip for only 1 or 2 seconds if the solution is fresh, longer if the solution has been weakened by long use. Almost immediately all areas affected with cupric oxide turn black. Unaffected areas remain a cream to light gray color. If left in the solution too long, gas bubbles quickly form on the work, indicating that the acid is biting into the silver itself. If allowed to continue, the acid etches away solder, dissolves nickel alloy findings, and leaves the work with a pitted surface difficult to remove.

After the cupric oxide has become black, thoroughly rinse the work in water, brush again with pumice until all the black is removed, and redip.

This may have to be repeated once or many times, depending on how often the work was heated and how high the temperatures were. When black cupric oxide no longer develops in the bright dip, and when all surfaces are a uniform light gray, the process is complete.

Since the surface of bright-dipped metal is slightly roughened even after careful application, it should be lightly sanded once more before going to the next step of patination (oxidation).

Safety Factors Involved in Pickling

Remember to stand well away from the pickle jar or vat when dropping in hot metal. A fine spray usually results which might affect clothing or skin long after the occurrence. Large work should be air-cooled first and then cleaned in hot pickle to avoid splashing.

Acid burns to skin or clothing should be immediately flushed with cold water and neutralized by a liberal dusting of sodium bicarbonate: If a skin burn is extensive, exclude air by covering it with a thick paste of sodium bicarbonate and water. Cover the area lightly with a gauze bandage and avoid greases or ointments.

APPLICATION OF FINDINGS

In most cases it is best to use sterling silver findings on sterling silver jewelry and the identical karat of gold finding for karat gold jewelry. Brass or nickel alloy findings cheapen the quality of the article as well as causing soldering and pickling difficulties.

Soldering Pin Assemblies

The fluxing and heat applications in this process are identical to ordinary hard soldering except that care should be exercised to prevent overheating of the projecting findings. This would cause the solder to run *into*, not around, the findings.

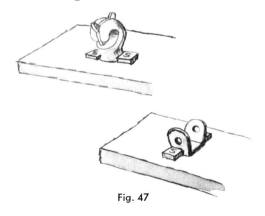

Fig. 47

50

The flame should be played around the findings until the solder begins to melt. Only then may the flame be concentrated on the findings themselves—and only for a few seconds. A figure-eight heating direction works well. (See Fig. 46.)

Solder may be pre-melted at the desired point or it may be applied around the base of the joint and catch in the manner shown in Fig. 47.

To prevent the solder from flowing into the holes of the joint and into the slide of the catch, it must be (1) placed flat and edging against the finding, (2) heated directly *only* after the base pieces are at the right temperature, or (3) protected by a touch of yellow ocher paste

Fig. 48

on the slide or in the holes of the joint. Be very sure that this paste does not mix with the flux!

The slide on the catch should be half open before heating is started. (See Fig. 48.)

The proper position of the joint and the catch is important to balance and appearance.

On a brooch worn with the long axis horizontal, the assembly should be placed above a median line so that the brooch will lie flat when worn. The catch should be to the right of the joint. The slot in the catch should face down so that, if the slide opens by accident, the weight of the brooch on the pin-

Fig. 50

stem will keep it in the catch. (See Fig. 49.)

When a brooch is to be worn vertically, the joint should be at the top, so that if the pin disengages from the catch the weight of the brooch will keep the pin in the cloth. The opening slot on the catch should face to the left.

For added spring to the pin, keeping it firmly in the catch slot even with the slide open, the joint should be slightly offset so that the pinstem extends behind the catch before insertion into the slot. (See Fig. 50.)

Use an "Easy" silver solder for ordinary work on silver and "Hard" on pieces to be enameled.

Soldering karat gold pin assemblies is identical to the above except that the appropriate karat gold solder is used.

Never solder with the pinstem in position. The high heat of soldering will anneal this nickel silver alloy so much that it will become very soft and will bend with the slightest pressure. Aligning the joint and the catch by eye works well enough and a tweezers or pointer should be in hand while soldering in order to reposition parts that move.

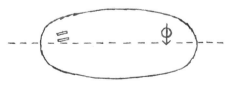

Fig. 49

Pinstems are made with a fixed rivet in place or with a hole through which a rivet is placed in attaching the pin to the joint. In both cases, pressure of some sort is used to spread the ends of the rivet to prevent its falling out. Special rivet-setting pliers are available which prevent the possibility of damage caused by rivet setting with a hammer.

Check the length of the pinstem to be used in placing the joint and the catch in position. Do not allow the point of the pinstem to project past the edge of the brooch. This is unsightly and might cause scratches during wear. The point of the pinstem should project no more than $\frac{1}{8}''$ past the catch.

Long pinstems may be shortened by clipping and refiling. If held in a ring clamp, the file marks may be removed easily and a sharp point formed by hard burnishing. All file marks must be removed or the point will not penetrate cloth easily.

Soldering Earring Backs

If hard solder were used in soldering earring backs, the delicate spring wires would become annealed and too soft for safe wear. Soft soldering is therefore necessary and is easy to do neatly if the solder is used economically.

Step 1. The spot to which the back is to be applied is touched with a bit of soft solder flux (glycerine and muriatic acid).

Step 2. Flux is also placed into the cup of the earring back.

Step 3. A small piece of soft solder ($\frac{3}{16}''$ long for the average cup) is placed into the cup and melted to fill it by use of a *small* soft flame.

Step 4. After waiting a moment, to allow the solder to cool, the finding is put in position.

Step 5. With the same small flame, the entire piece is heated only enough to allow the solder to show around the edge of the cup. Excess or prolonged heat will cause the soft solder to eat into the silver. This causes a rough depression impossible to remove.

Step 6. After the solder has melted, do not immediately cool the work. The solder remains fluid for some time. Cooling may be speeded by dropping water from a fingertip onto the join.

Step 7. After the solder has hardened, the piece may be cooled in water. It is useless to pickle soft soldered work since the flux coats everything with an acid-proof film.

Step 8. The piece should be thoroughly scrubbed with soap and pumice to remove flux. Even better, it should be boiled out in a sodium bicarbonate solution. If some of the muriatic acid remains around the finding, severe burning of the ear lobe could result.

Step 9. Any soft solder "ghosts" should be removed with a scraper and scotch stone.

Step 10. Work may now be given a patina and finished.

Remember! No hard soldering may be done *after* work has been soft soldered.

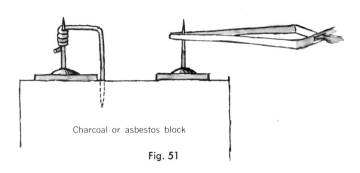

Charcoal or asbestos block

Fig. 51

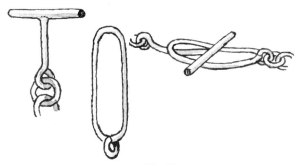

Fig. 52

Soldering Tie Tacks

The same techniques as for soldering earring backs apply for soldering the post of a tie tack to the back of the designed form. Here again complete neutralization of acids is necessary to prevent damage to cloth and skin.

The tie tack may be held in position with locking tweezers or with iron binding wire. (See Fig. 51.)

Fastening Spring Rings and Jump Rings to Chain

With commercial findings of this sort, no soldering is used since high heat would affect the strength of the parts. On small chains the end links may be enlarged by inserting a pointed scribe into the last link and pushing into a soft wood block. The small ring on a spring ring is opened by a slight twist, the chain link inserted, and the ring again tightly closed. Two chain pliers are used for this.

Many craftsmen prefer to design their own findings for necklaces. In most cases these are better integrated with the chain design itself. A very basic form, the toggle, may be varied to adapt it to many types of chain. (See Fig. 52.)

COLORING (OXIDATION)

The term "oxidation," though often used in describing the planned coloring of silver and other metals, is a misnomer. The metal does not oxidize but rather is affected by several sulfur compounds to form metal sulfides. The term "coloring" which covers the chemical color change of many metals in many ways is more accurate.

The work may be colored only after all soldering, both hard or soft, has been completed and after the work has been thoroughly cleaned of fire oxides by pickling.

The chemicals used may vary considerably, but the method of their application is often identical.

The work must be clean and free of oil or grease. With a brush and pumice remove all of the pure silver surface film left after pickling. Then immerse the work in the coloring solution until the desired strength of color is reached. By rubbing the surfaces easily reached with fine pumice, one may create highlights which enrich both the form and the textural quality of the work.

It is best to remove the color only from places easy to reach since protected areas freed of color will tend to tarnish again quickly. On the other hand, it is not practical to allow color to remain where the friction of use will gradually rub it off.

It is best to use the thumb and fingers for developing highlights. Brushes, emery paper, or steel wool tend to remove too much of the color, as well as scratching areas best left dark.

After pumice rubbing, flush out hidden grains of pumice with a sharp jet of water. If allowed to remain, they would contaminate the final polishing medium, causing scratches.

Coloring Techniques for Silver

Dark Gray

Solution: Potassium sulfide
(liver of sulfur, K_2S) .... ½" cube
Water .... 2 pts.

The liver of sulfur dissolves quickly if heated almost to boiling temperature.

The dry lump chemical as well as the above solution must be kept in dark, air-tight bottles to prevent deterioration.

1. Attach a length of silver wire to the cleaned silver piece and dip it completely into the heated—but not boiling—solution.

2. Remove the work quickly to check progress. If the solution is too hot, a heavy coating of silver sulfide is deposited which chips and flakes off when rubbed. The desired application creates a smooth transition from dark areas to light areas while rubbing. If the application is too heavy, it must all be removed by brushing or reheating and pickling. Then it may again be carefully dipped into the solution.

3. If the color is mottled with blues, purples, and greens, dip the work again and again until a uniform gray-black is obtained. Never leave the work in the solution while attending to something else!

4. Quickly rinse off all traces of the solution; otherwise it will continue to affect the metal.

5. Rub off excess sulfide with fine pumice. This should be moist, not wet. Moisten the thumb or forefinger in water, then dip it into the dry pumice before rubbing. If too wet, the pumice runs off the work without effective cutting. Rub in all directions to avoid parallel scratches.

6. Rinse the work thoroughly, dry it completely and polish—preferably by hand since wheel polishing tends to remove much of the coloring from places where it should remain.

7. A light application of vegetable oil darkens the oxidation considerably. The work should be wiped dry afterward or the oil will catch and hold dust. A very thin film of beeswax dissolved in benzol works well and also protects the polished areas from tarnish for some time.

Coloring chain with liver of sulfur or other chemicals is often necessary to reduce its often garish brightness as it comes from the manufacturer. A very bright silver chain attached to a more subtly colored pendant calls too much attention to itself, where the pendant should be the focal point of interest.

To color chain, the whole length of chain is simply immersed in the warm coloring solution long enough for a solid color to develop. The spring ring—the small ring with an enclosed, spring-activated slide—is left off until after the chain has been both colored and polished, for the pumice removal of excess color would clog the spring, thus ruining its action.

Excess color is removed by rubbing down the length of the chain with a pinch of moist pumice held between the thumb and the first two fingers. Rub down a number of times in both directions to remove an adequate amount of the color.

GOLD COLOR

Solution: Ammonium sulfide
 (NH_4S) .... 1 Gm.
 Water .... 7 oz.

1. Use the solution cold.

2. Dip the work and remove it when the desired shade is reached. A variety of shades from crimson to purple and brown may be achieved by repeated immersion or by heating the solution.

3. Remove color where desired as in the first formula.

LIGHT GRAY

Solution: Platinum chloride $(PtCl_4)$ in alcohol

1. Work may be dipped or the solution may be painted on.

2. A variety of gray to black values may be reached by timing the dipping.

3. Finish as in the first formula.

BLUE COLOR

Place the work in a closed steel box with a little pure sulfur (S). Heat together until the silver turns blue. Do not allow the sulfur to touch the silver directly. For an even coloring place a little sulfur in each corner of the box.

DEAD BLACK

Solution: A concentration of ammonium sulfide (NH_4S) in water used hot

Use in the same way as the first formula, avoiding an over-application.

Coloring Techniques for Gold

BLACK ON 14K ALLOYS CONTAINING COPPER

Solution: Liver of sulfur as for silver, but used hot and with the gold heated before dipping

PURE GOLD COLOR ON KARAT GOLDS

A pure gold surface on karat alloys may be achieved by combining the following in a heated crucible:

Potassium nitrate (KNO_3) ...... 2 parts
Common salt .... 1 part
Alum .... 1 part

When heated, this combination becomes fluid. The work is first dipped into an acid solution of 1 part nitric acid to 10 parts water, rinsed in boiling water, and agitated in the crucible solution for a few minutes. After rinsing the work each time in boiling water, it may be dipped again and again until the color of pure gold is reached. This surface film is thin, so care must be exercised in polishing.

All the colors of gold (such as pink, antique, yellow, red, white, green, etc.) may be achieved by electroplating the entire work or separate areas not protected by a resistant film. Though plating is a widespread practice, and has been used in one form or another for centuries, it would seem best to exploit the worked material to its best advantage rather than to disguise it with something else.

Coloring Techniques for Copper and Brass

BROWN COLOR

Solution: Copper sulfate ... 1 part
            Water .... 2 parts

Dip work in a hot solution. Prolonged dipping forms a darker brown. For the most even coating it is best to brush the work thoroughly with pumice and water after the first brown coating. Reapplication of the above solution develops a more uniform deposit.

BLACK COLOR

Solution: Barium sulfide or ammonium sulfide used as a hot concentrated solution. Dip carefully to avoid over-application.

Copper and metals containing copper develop green and blue surface coatings when long exposed to air or earth containing certain chemicals. Most sculpture using copper, brass, or bronze is given this patina upon completion.

The Japanese metal craftsmen have for centuries used patinas on a great variety of alloys and have developed this surface coloring to a rich and highly perfected art.

Perhaps the richest patina for copper alloy metals is the green formed by copper nitrate. By changing the formula and the procedure, sage green, olive green, blue green, and a rich dark green may be achieved.

SAGE GREEN

Solution: Copper nitrate .. 1½ Gm.
            Water .... 6 oz.

Use the solution hot and paint it on the chemically clean metal. Several applications may be necessary.

OLIVE GREEN

Solution: Iron perchloride or ammonium chloride .... 1 part
            Water .... 2 parts

Apply as in the above formula.

DARK GREEN

Solution: A paste is formed of

Copper sulfate	....	1 part
Zinc chloride	.....	1 part
Water	....	1 part

Apply the paste to the metal, allowing it to dry completely. Wash off the paste and expose the work to sunlight.

ANTIQUE GREEN FOR BRONZE

H. Wilson gives this Japanese formula:

Solution: Copper nitrate	48 grains
Sal-ammonia (ammonium chloride) ...	48 grains
Calcium chloride	20 grains
Copper sulfate	10 grains
Oxalic acid ...	10 grains
Water	4 fluid oz.

Additional ammonium chloride and copper sulfate create a darker color. A bright green patina results when the copper sulfate and the oxalic acid are omitted.

The clean metal is given a coating of this solution each day for several days. When the desired color is reached, the surface should be brushed with a dry brush. This may be done for several days, after which the color may be fixed by applying a thin, even coat of beeswax.

Beeswax may be brought to brushing consistency by heat or by dissolving it in benzol. The benzol evaporates, leaving a thin film of wax. Since this darkens the patina, an experiment should be made on a scrap of the same alloy before wax application.

Sometimes bronze develops a corrosive "disease" which continues to eat into the healthy metal until it has all disintegrated to a dry white powder; the beeswax film helps to prevent this.

BLACK PATINA FOR BRONZE

Solution: Ammonium sulfate		1 part
Water	....	2 parts

The solution is brushed onto the warmed, clean metal. Allow the surface to dry. Rinse the object in warm water and repeat the brushing with the solution until the desired color is obtained.

POLISHING BY HAND AND MACHINE METHODS

The final basic step in jewelry construction is the surface finishing. Some designs lend themselves best to a highly reflective finish while others will require a matte quality. Generally, forms which have contour—positives or negatives—can best use the high polish. In forms with contour, the scratches of wear and the dulling through handling are kept to a minimum since only small portions of the metal present a surface for contact. In addition, the play of reflected light and images over a polished, undulating surface is rich and varied. This quality increases the illusion of volume and lends a fluid softness to an otherwise rigid and resistant metal.

Flat planes form a surface which is easily affected on its entire area by a scratch or by fingerprints. A surface which is already textured by fine lines or fine scratches disguises many of these blemishes.

Jewelry, if worn often, acquires a surface quality of its own in time which must be considered characteristic and therefore quite legitimate.

The practice of applying a film of lacquer or plastic to metal is, in the author's estimation, unsound. Such a film protects a high polish for a short while only and soon becomes yellowed or spotty. To remove it and apply a new coat is a poor use of time when a little rubbing with a rouge-impregnated flannel cloth restores the *metallic* quality much more easily.

Lacquered metal loses the feeling of metal, its temperature, its hardness, and its precision of edge and surface.

Lacquering would be justified only when work must be displayed for lengths of time in exhibitions. Here it is preferable to having someone experienced with polishing do the occasional necessary cleaning.

Hand Polishing

Polishing by hand has the advantage of safety and, unless one is experienced in using polishing motors, precision.

If, throughout the construction processes, all scratches, dents, and warping are avoided or corrected, a final high polish is easy to apply by hand.

1. Remove all traces of coarser abrasives (such as fine pumice used in the coloring process). Flush the work under a hard jet of water and boil it in a solution of 1 tbsp. liquid detergent to 1 pt. water. Brush out the dissolved residue with a soft brush.

2. Dry the work thoroughly. Moisture in crevices and along wire appliqué will collect rouge dust which might be difficult to remove later. Blotting the work with cleansing tissue does this efficiently and well.

3. Examine the surface. If it seems quite dull, it should be buffed with a felt buffing stick to which Lea compound or tripoli has been applied. Boil out the work again as in Step 1 and dry it thoroughly.

4. When dry, the clean metal areas—not those left colored—are buffed with another felt-covered stick to which rouge has been applied. Rub rapidly with as much pressure as the work will allow. Change the direction of rubbing constantly to obtain a uniform polish.

5. It is usually at this point that the cupric oxide deposit begins to show itself if it has not already been completely removed by bright dipping. Since a high polish is impossible over such a deposit, it is necessary to go back to this step. Of course the metal must again be recolored, since this is also removed in bright dipping.

6. Once the polish is obtained, the work should again be boiled in a detergent and water solution.

A final light buffing with rouge (not enough to deposit dust again) spreads a protecting film of wax (used as a binder in stick rouge) on the metal.

A rouge-impregnated flannel cloth is used to brighten the highlights of a chain. Hold one end of the chain and rub down its length between two layers of cloth with the other hand. Reverse the direction a few times.

Trying to polish areas that are difficult to reach is wasted effort since they will tarnish quickly again in normal use.

Red rouge is used for silver and for red, green, or yellow golds.

White rouge is used for the platinum metals and for white gold.

Black and green rouge are used on all the above metals for a brighter polish.

Areas that are impossible to reach by hand buffing may be polished by burnishing. A straight or curved burnisher is used applied with pressure and with a lubricant of a soda-free paste soap or light oil. Rouge or diamantine in cake form may be used as an additional fine abrasive while burnishing.

The burnisher should occasionally be polished on a piece of leather during the process and should be coated with a film of beeswax and wrapped in chamois when not in use.

Small or delicate areas may be polished by trumming with a cord; cake or stick rouge is first rubbed into the cord.

Matte Finishes by Hand

Matte finishes may be achieved without a motor in several ways:

FROSTING

The completed work, after having been pickled and bright dipped, is dropped hot into pickle, thus dissolving the copper oxides at the surface and leaving a light layer of the pure metal.

A solution of 1 part sulfuric acid to 8 parts water is used with sterling silver while the same proportions of nitric acid and water may be used on karat golds containing copper. Several heatings and quenchings may be needed to achieve uniformity, and boiling the work in the pickle each time speeds the process. Of course coloring the metal and consequent highlighting are impossible since the fine film of pure metal would also be removed. A careful hand rubbing with whiting and water may be used to pick out highlights here and there, but avoid using an abrasive which must eventually be scrubbed clean.

SAND BLASTING

A frosted surface may be given to any metal by blowing at it a fine, sharp grit with compressed air. The necessary equipment might be expensive but a usable blower might be made of a large airbrush unit. Use more than one grit size to reduce clogging of feed pipes. The grit may be carborundum, stone powder, sand, or quartz.

Areas that must be protected from the abrasive may be painted with hot wax or rubber cement. They may also be masked out with pieces of electrician's plastic tape cut to shape for very precise forms. The grit is either caught or deflected in this way.

By changing grits and air pressure, a considerable variety of surfaces may be achieved.

STEEL WOOL

The work may be rubbed, always *in one direction only*, with fine steel wool to obtain a bright matte finish. This works well on flat, exposed surfaces and allows partial highlighting of curves and edges. The steel wool should be folded into a compact ball and used with considerable pressure. Steel wool sizes from 1 down to 000 are recommended. If larger than 1, the scratches no longer

appear as a fine surface texture; if finer than 000, the surface loses the matte quality desired.

Machine Polishing

Polishing to a high finish by machine is a time-saving process but requires greater skill and experience than polishing by hand. A motor with an rpm rating of at least 1800 is necessary for efficient polishing. The best quality, close-weave canton-flannel wheel should be used. With an 1800 rpm motor, a wheel of at least 11″ is necessary to obtain the high speed in surface feet needed. If a smaller wheel is used, such as 6″ or 7″ diameter, an rpm of 3000 is necessary.

The heat of friction, when the correct abrasive (rouge) and the correct speed are used, causes the surface of the work to flow. Minute scratches and pits left by previous operations are thus filled in and smoothed over. Consequently it is important to "cross-polish." This means that the work is constantly moved into a new polishing position during the operation.

In order to *charge* the wheel with rouge effectively, the rouge bar should first be dipped in kerosene. This allows the rouge to penetrate the wheel fibers and less is blown away during the process.

When rouge cakes on the surface of the work during polishing, it means that too much rouge has been used or that the wheel speed is too slow.

Do not apply excessive pressure—this will not give a better polish but instead will "burn" the metal, causing roughness. A firm, even pressure throughout the operation is sufficient.

When wheel-polishing delicate sheet or wire forms, it is necessary to support the work firmly on a leather-covered board.

It is very dangerous to machine-polish chain without safety precautions.

Chain should be wrapped firmly around a board with rounded edges. The ends of the chain may be pinned down securely with tacks. On very delicate chain, apply a lighter than usual pressure to prevent stretching the links.

Hold all work with cleansing tissue. Fingerprints may also cause rouge dust to cake on the metal surface.

Hard felt or wood wheels are available in a variety of shapes and sizes for *lap polishing*. This technique is used where the sharpness of angles and edges must be preserved. An occasional reshaping of these wheels is quite simple, using the edge of a file or hacksaw blade while the wheel is in motion. Avoid cutting grooves into the wheel in this manner.

A nylon bristle wheel, using a paste of whiting and water, polishes delicate wire and filigree safely.

Matte Finishes by Machine

PUMICE AND MUSLIN BUFF

Remove all blemishes with emery paper or by wheel cutting. Give the work a uniform satin finish using a 6″ muslin wheel revolving at 1700 rpm and a paste made of pumice and water. Wash work thoroughly and color it. Polish again with the same materials to bring out highlights. Wash again and dry the work with a clean soft cloth.

WIRE BRUSHING

Use a wheel of nickel wire, fiber or nylon bristles, and a lubricating solution mixed with fine pumice. Wheel speeds for this purpose are much slower than for cutting or bright polishing. Speeds of 600 to 1200 rpm are sufficient. Reversing the direction of a wire wheel now and then prevents bent tips which will no longer scratch evenly. Use very light pressure since bent tips flail the surface rather than texturing it. The lubricating solution may be kerosene or soft soap and water, mixed to a milky consistency.

The commercial polishing techniques involving electropolishing, drum tumbling, and other production processes are usually not appropriate for use by the hand craftsmen, being designed for volume production.

3 · SUPPLEMENTARY METAL TECHNIQUES

Though an imaginative designer may successfully limit himself to the quite basic construction techniques already described, there soon comes a time when the need for greater expressive latitude becomes felt.

Such more specialized processes as forming, repoussé and chasing, casting, and so forth, introduce a new dimension to the jewelry form. So far we have found that the basic forms of metal (sheets—rigid and usually quite thin—as well as wire) lend themselves very well to a two-dimensional, almost graphic development. The translation of a visual idea to a sketch and further into a cut-out metal shape is quite direct.

However, to exploit the full richness of metal, its reflective qualities and its malleability, a three-dimensional development may become necessary. This is achieved in two ways. Metal—in its sheet or wire basic form—may be bent or stretched to create a contour of positive or negative shape, or, having been heated to its melting point, it may be cast as a liquid into a designed form in a mold.

These and other techniques will be described in this chapter.

FORMING

The same techniques and requirements are used in forming jewelry shapes as are employed in silversmithing, though on a much smaller scale.

By forming, giving contour to flat metal, a depth of dimension is created that results in a very different quality from work using only flat planes. The play of reflections and light over an undulating and highly polished surface lends a sculptural quality which often transcends the small size of the total form.

Though any practical gauge of metal may be formed, the structural strength achieved by forming allows the use of somewhat thinner gauges than necessary for flat designs.

The aesthetic danger in thin gauges lies in the appearance of edges. When these are thin, a feeling of flimsy lightness results which seriously detracts from the basic solidity of the work. A thickened or reinforced edge has been used by silversmiths for centuries to correct this disadvantage.

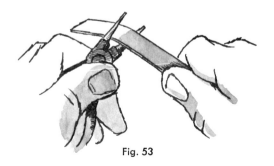

Fig. 53

60

Fig. 54

In most cases, a really extreme stretching of metal in forming is not necessary in jewelry making, so metal of 18 or 20 gauge, B and S, is sufficient. Forming techniques are basically of two types: bending and depressing. In bending the flat metal is left in its original thickness throughout by merely twisting or curving it with pliers or over mandrels. It may also be curved by forcing it into half cylindrical grooves or other shapes in wood or steel.

Pliers for bending should have absolutely smooth, polished jaws. Jaws with serrations cause dents and roughness difficult to remove.

Do not hold the metal in the jaws tightly. Use a round-nose pliers for curved shapes and a flat or chain pliers for angles. One jaw is used as a fulcrum around which the metal is carefully bent. The little finger of the pliers-holding hand is used as a brace to maintain the correct opening. This technique prevents denting flat metal as well as wire.

Special bending pliers have one flat jaw and one round or curved jaw. When using such pliers, always keep the flat jaw on the *outside* of the curve, for the sharp edges would mar an inside curve. (See Fig. 55.)

Forming by depressing uses punches or hammers to gradually force flat metal into a positive or negative contour so that the total surface is affected. The most direct method, and one in which few hammer marks are left, is to use a preshaped depression in wood or lead into which the metal is hammered, preferably with a punch shaped to fit the depression.

If the depression is carved into a maple or birch block, and hardwood punches are used, many shapes of the

same dimension may be made from the same die. Lead, being softer, loses its shape, and a mold shape may be used only once.

Where greater flexibility is needed—i.e., where a shape becomes complex in contour—the metal is shaped on a lead block with light blows of a hammer. These are the basic steps:

1. The metal is first *annealed* by heating and quenching. Metals become hardened when compressed or bent. Rolling, hammering, and stretching change the basic structure so that at a point the metal cracks. This "work hardening" may be corrected by heating the metal and, depending on its chemistry, cooling it quickly after heat has expanded it or allowing it to cool slowly in the air. If properly done, the metal is then ready for additional working.

Sterling silver is heated to 1100° F and quenched immediately in cold water or pickle. Be sure to have the water or pickle close to the annealing area to prevent cooling. Handy Paste Flux may be used as an accurate temperature in-

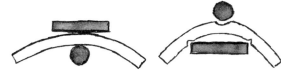

Fig. 55

dicator since it turns to a water-clear glaze at 1100° F. The metal, if viewed in a darkened area, should show the first dull red glow.

Yellow and green golds are heated to 1400° F, or cherry red, and may be air-cooled or quenched.

Red golds, also heated to 1200° F, should be quenched red hot to avoid rehardening.

White golds are best annealed at 1400°F, or cherry red, and may be quenched or air-cooled.

In most cases, sheet stock comes already annealed, but it may often be further softened by annealing prior to working.

2. It is best to cut the shape to be formed somewhat larger than the final shape should be. In depressing the metal the outer edges pull in as the contour increases. Do not worry about edges at this point beyond filing away rough saw marks.

The final trueing of edges and refinement of the surfaces must be done after forming is completed.

3. The sheet is placed on a smooth lead block or over a depression already hammered into the block. A lead block may be made by pouring molten lead into a small pan such as those used in schools for tempera paint. Oil the inside of the pan lightly so that the lead will not stick. Be sure that there is no water in the pan or the hot lead will spatter. A block which has become too pitted or warped may easily be recast in the same way. Lead may be melted on a gas burner in any iron vessel. Brush the surface slag to one side with a stick of wood just before pouring.

4. The hammer should have a polished dome or sphere at one end and a polished broad surface at the other. A *french chasing hammer* is ideal for this purpose since it is small, light, and easily controlled.

5. Determine which area is to be depressed most. Start hammering with light, evenly weighted blows. Develop the shape slowly, working outward in a tight spiral from a central point. The weight of the hand and the hammer

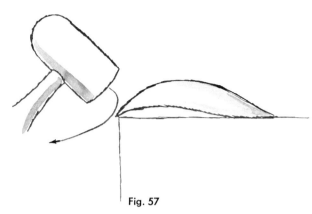

Fig. 57

should supply enough downward force. Keep the blows so close together that one overlaps the other. In this way the growth of the contour is even in shape and texture. (See Fig. 56.)

6. Hammer to within $\frac{1}{8}''$ of the edge since this should be left at its original thickness.

7. After the first *course* of hammering, the piece may have been warped out of the desired shape. Before annealing, true the shape by hammering it on a wooden surface with a wood mallet. If edges are to remain on the same plane, they may be forced down by glancing blows of a wood or plastic mallet while the work is held at the edge of the bench. (See Fig. 57.)

8. The metal should be annealed after each course if it is to be deeply depressed. Important: when working on a lead block all flakes and particles of lead *must* be removed before annealing. Even very small amounts of lead eat into and destroy the surface of precious metals when heated above 600° F. Lead may be removed by a thorough rubbing with steel wool or emery paper.

9. Start each new course from the same point and hammer uniformly over the entire depression to avoid hollows and dips. The piece may be angled to accommodate a new direction of hammering, and much warping may be avoided in this way.

10. After each course—and before annealing—the edge should be thickened

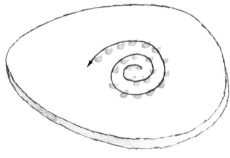

Fig. 56

by hammering. If cracks should develop along the edge, clean and flux the work and solder the crack with *Hard* solder. An untended crack will work deeper into the form with each course.

11. After the basic contour is achieved, the hammer marks must be removed. In most cases they are too raw and out of scale to be used as texture in jewelry. If the blows have been small and uniform, the refinement may be done with files and emery paper. A riffler file may be used on concave areas and a large flat hand file for the convex surface and the edges. Sanding will remove file marks or a scotch stone might be ground to shape for stoning concave areas. Finishing may then proceed as usual.

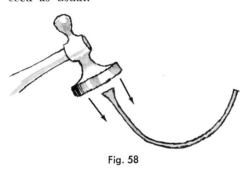

Fig. 58

12. On simple forms the hammer blows may be removed by planishing. Use a polished steel form—a *stake*—which best follows the inner contour of the metal shape. Using the flat end of the planishing hammer, work again from the center toward the outside edges in an even, *growing* manner. Use lighter strokes than for forming. A slightly circular glancing blow helps fuse one plane into the other. (See Fig. 59.)

Overlap each blow and hit *only* where the metal is completely in contact with the stake. This contact may be felt by the solidity of the blow. A blow over air space feels very different.

Avoid thinning the metal more than necessary. Hammered over steel, the metal stretches horizontally and may

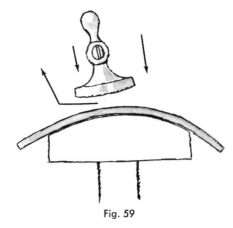

Fig. 59

warp badly if the blows are uneven or too heavy. Avoid planishing to the edge. This should be left as thick as possible.

If done correctly, planishing smoothes away dents on the inside at the same time the outside is refined. A little sanding or cutting on the wheel is all that is necessary to develop a smooth curve, free of facets and indentations.

If possible, the work should not be annealed or soldered again. Planishing has work-hardened the metal to good advantage on thin forms. If soldering must be done, allow the work to air-cool completely and pickle it by boiling or soaking. Do not quench it hot!

13. Sharp angles or changes of direction are possible using the correct stakes. After the desired degrees of contour are formed, the metal may be bent over the slightly rounded edge of a polished, flat anvil stake. Once a bend is started, be sure to place the edge of the stake in the same crease each time. Serious scratches and dents can result if this is done carelessly. (See Fig. 60.)

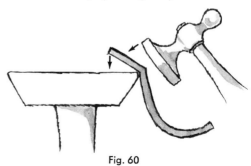

Fig. 60

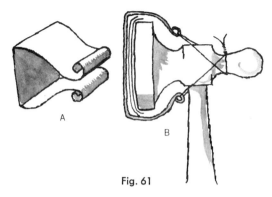

Fig. 61

Do not try to form the angle with the first hammer blows. Let the new plane develop slowly with light, even, and *parallel* strokes. Edging the hammer forms dents almost impossible to remove completely.

Keep all hammers, stakes, and anvils rust-free and polished. After use a light film of oil should always be applied which may be wiped off to avoid contaminating the work before annealing.

A hammer may be constructed which leaves no mark on the surface of the metal during planishing. A strip of spring steel is formed, as shown in Fig. 61A. This is fastened to the hammer with heavy iron wire, first placing a cushion of fine textured cloth and paper over the hammer face. Fasten this buffer tightly enough so that it does not slip about in hammering (Fig. 61B).

REPOUSSÉ AND CHASING

Though there is a basic difference between the two, most work that uses repoussé must use chasing as well. Repoussé is simply the forming of a low or high relief by working sheet metal from

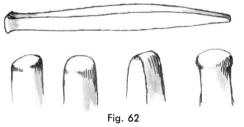

Fig. 62

the back. Chasing consists of defining, delineating, and texturing the surface of sheet metal after repoussé relief is complete, or on flat surfaces alone.

The tools and their uses are very similar. A chasing hammer is used for small work and any larger hammer with a broad flat face may be used for large work.

Various punches may be made or purchased, each fulfilling a function. For repoussé, punches of a variety of shapes and diameters are used for forcing up

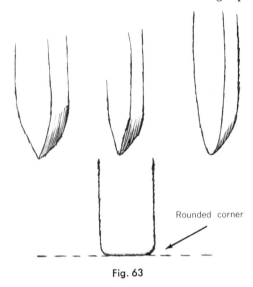

Rounded corner

Fig. 63

the metal from behind. These are made of tool steel, are free of pits, and should be highly polished. Most repoussé tools are about 5″ long, are square or rectangular in section, and may be tapered at both ends to give a firmer gripping area in the middle. (See Fig. 62.)

Chasing punches are made in a much greater variety of shapes—as many as 70 are commonly used. They fall into a few use categories:

1. *Tracers*—These are shaped basically like blunted chisels. They may have very thin edges or broad, depending on the quality of line they should form. It is best that they be slightly rounded on the corners to prevent cutting the metal. (See Fig. 63.)

Fig. 64

2. *Curve punches*—These are usually used with one blow of the hammer to incise the shape of the tool. They may vary from flat semicircles to almost full round. Others may form a variety of angular shapes. (See Fig. 64.)

3. *Modeling tools*—These are flat or convex tipped tools for working the surface smooth or for punching down areas around raised relief. Be sure that the edges of the face are slightly rounded. (See Fig. 65.)

are used to punch circular lines as texture or outline. They may be purchased in several diameters. (See Fig. 67.)

The craftsman is seldom able to purchase all of the repoussé and chasing tools he needs. They may be made by purchasing forged blanks which may need only to have the ends shaped or by using square, flat, or round tool steel stock.

The stock should not be thicker than $\frac{1}{4}''$; much larger stock is difficult to

Fig. 65

4. *Matting and graining tools*—These are used to texture a contour or to enrich the background around relief areas. They may have an infinite number of faces, each giving a specific texture indentation.

Fig. 66F may be made by filing a notch around a length of square or round tool steel, breaking it off, and using the rough end as the texture surface.

5. *Ring tools*—Made of round stock with a hemispherical depression in the tip, these tools must be well tempered and hardened since they may be easily dulled or have their edges broken. They

hold and control. Taper the rod by forging or filing both ends. This makes observation of the tool mark easier.

The working face may be filed to shape by use of needle files—first a coarse file and then a smooth worn file—and then polished with gradually finer degrees of emery cloth or paper. The corners of square or flat stock should be beveled to an octagon shape for easy holding, and the sides may be roughened to prevent slipping by *drawing* the flat file along the length at right angles to the stock. (See Fig. 68.)

To harden the working face, the last $\frac{1}{2}''$ should be heated evenly to a bright

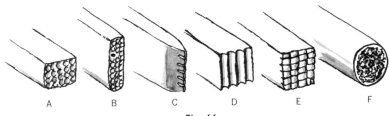

A   B   C   D   E   F

Fig. 66

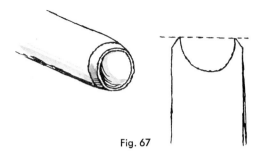

Fig. 67

red. Quench the tip very quickly in cold water, which should be in a container next to the heating area.

Polish the working face brightly once more and heat it again in a fairly soft flame at a point about 1″ from the face. The heat will travel down toward the face, changing the color of the steel. The very moment that the face turns a dark yellow it must again be quenched.

Many texture faces may be made in soft tool stock which may give a personal and very individual quality to one's work. (See Fig. 69.)

The Repoussé Technique

For repoussé, the work must be held on a surface which is both firm and resilient. A pitch mixture consisting of burgundy pitch, tallow, and plaster of paris with a little linseed oil added has been used for this purpose for centuries. Very good pitch mixtures may be purchased ready mixed, but some craftsmen prefer to develop their own degrees of hardness or softness.

A basic formula for pitch is:

Burgundy or Swedish pitch .... 7 parts
Plaster of paris or powdered pumice 10 parts
Tallow or linseed oil .... 1 part

More linseed oil may be added in winter when the pitch is cold, less in summer when more firmness is needed.

The prepared pitch is heated and poured into a form. This might be a pitch bowl held in a ring, or a shallow wooden box with edges 1½″ to 2″ high. The pitch should be poured so that it is higher in the center of the container than at the sides.

Melted pitch may also be poured into a hollow object which may then be worked from the outside without danger of excessive denting.

The procedures in repoussé are:

1. Transfer the drawing in reverse to the cleaned reverse side of the metal with pencil or carbon paper. Lightly scratch the lines for permanence.

2. Heat the pitch so that a flat, smooth, and bubble-free surface is formed. Do not permit the pitch to burn. The hard ashes of burned pitch give a rough backing to a working area that will cause difficulties. If a mound is needed to support a contoured piece, allow the pitch to cool a bit, wet the fingers, and model it to shape. The surface must again be heated to slight flowing before the work is applied.

3. Heat the work slightly and place it on the warmed pitch. Do not press it in too deeply or the pitch will flow over and hide the edges. Press down enough to form an even contact between the metal and the pitch.

4. When the pitch is quite firm, work a little over the edge of the piece to key it in. Allow it to cool completely. A piece may disappear very quickly when hammered into warm pitch!

5. Holding the tool as shown in Fig. 71, punch up the high relief gradually. Work from the lowest areas to the highest point, annealing as often as necessary. As soon as the metal resists the blow of the punch it should be annealed, since continued punching will only loosen the metal from the pitch, warp the entire shape, or even cause the overstretched metal to crack.

6. Removing the work from the pitch surface may be done in several ways:

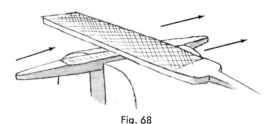

Fig. 68

66

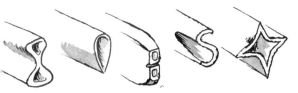

Fig. 69

(a) The bottom of the pitch container may be given a sharp blow with a mallet, often dislodging small pieces of work which may already be loose. (b) A tracing tool worked between the metal and the pitch might pry it away with slight pressure. (c) The metal may be warmed by a small flame and lifted off.

When melted and soft, pitch may be wiped off of metal, especially if a cloth soaked in turpentine or benzol is used. It may also be burned off during the annealing process. Make sure that all traces of pitch have turned to white ash. A partially burned area of pitch, if dropped in pickle, will form a hard crust difficult to remove.

7. After annealing and pickling, dry the work and replace it in the pitch, again making sure that it is evenly supported over its entire surface.

8. Having observed areas that need greater contour, you may now work these to completion.

9. Remove the work, anneal, pickle and dry it, and replace it on the pitch with the top up for necessary forming or chasing on this surface. Avoid indenting the background more than necessary. Once the areas to remain flat are stretched, it is difficult or impossible to flatten what is then excess metal.

10. To flatten or model the back-

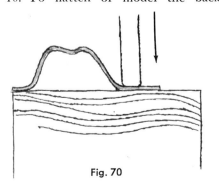

Fig. 70

ground, the cleaned work is placed on a block of close-grained wood. Use modeling tools to depress and flatten areas adjacent to the raised relief. (See Fig. 70.)

11. The proper action of the tool is to cause it to move slightly with each hammer blow so that one mark flows evenly into the next. Do not lift and replace the tool for each blow. The hammer blows should be light and made as rapidly as possible in order to move the tool evenly.

The modeling tool, as well as other chasing tools, should be held as illus-

Fig. 71

trated in Fig. 71. The thumb and first and second fingers hold the tool while the third and fourth fingers are braced on the working surface. (See Fig. 71.)

12. When textured tools are used, each position should be planned and the tool carefully placed in position before the hammer blow is struck. Indiscriminate hammering will cause a poorly defined surface.

13. Blunt relief forms may be sharpened and defined with flat modeling tools. It is best to melt pieces of pitch into the back of these areas first to support the metal. (See Fig. 72.)

14. If done skillfully, the surface created by the tools needs no additional finishing. If scratches or dents must be removed, use a scotch stone and water or a boxwood stick with pumice and oil. Try to anticipate the effect of coloring the metal when determining the propor-

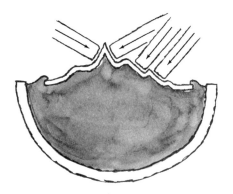

Fig. 72

tion of textured to smooth surfaces. Remember that rough or textured areas hold color better and remain darker than smooth areas. If highlights are needed, buff them up with a hand buff and tripoli before coloring and polishing.

Chasing

The chasing process uses many of the previous tools and techniques, but instead of using a resilient surface most chasing is done on a block of hardwood or steel. The work may be pegged down by careful placement of headless nails. For additional safety the nails might be covered with small lengths of plastic tubing. Since the force of hammering moves both the chasing tool and the metal, these pegs should act as stops rather than as clamps. Often the same position of the pegs may be used for a great variety of shapes and sizes of metal. (See Fig. 73.)

The technique of chasing a straight or curved line is as follows:

1. The design is *lightly* scratched onto the metal and the work is fixed to the wood block.

2. Holding the tool as shown in Fig. 71, place it at the start of a line. Whereas in modeling or punching the

tool is held perpendicular to the work surface, the tool is now angled slightly. You can see this depicted in Fig. 74. The correct angle causes the tool to move toward the worker with each blow. In using the corner rather than the whole edge of a tracing tool, an even line is formed by continuous rapid blows of the chasing hammer. About 140 blows per minute is an average speed. If the total edge is used, a series of short jagged depressions results. This is very true when a curve is formed. If well done, no *stitches* should show on the edges or the bottom of the chased line.

The wood block should be heavy enough so that it does not move with each hammer blow. It may be clamped into a vise or to the bench with a "C" clamp, but this limits easy movement as curves are chased.

Other surfaces suitable for chasing are a steel block, pitch, leather with the rough side up and wetted, and a lead block. All but the first surface mentioned cause the chased line to show as an indentation on the back of the piece if it is 16 gauge B and S or thinner.

Chasing on a pitch bowl has the advantage that one may maintain the best tool angle while hammering vertically. The bowl, in its ring, is tilted to the best plane. Even gravity seems to help in this. Dip the ends of chasing tools in a light oil for smoother working.

It is difficult to correct a slip with the chasing tool because the metal has been hardened around the line. Careful planishing or burnishing might help. Sometimes a slip may be remedied by engraving over the chased line, but this always leaves a very different effect.

Fig. 73

CASTING

There are many methods by which metals may be cast, some of them as primitive and limited today as they were thousands of years ago when first developed. The basic principle is the same: molten metal is poured into a mold.

The earliest examples of molds are of baked clay or soft porous stone. Even today the Indians of the Southwest use a form of *tufa,* a light volcanic stone, into which a design is carved. The clay mold, with varying details of construction, was developed independently on all continents, Africa, the Far East, India, and South and Central America.

In most cases the form to be cast was first modeled in wax. The wax was carefully enclosed in clay, with an opening through which it could be melted out and into which the molten metal could be poured. This is the almost universal *cire-perdue* or *lost wax process.*

Around the beginning of this century a method was developed for forcing metal into a mold with great pressure by centrifugal action. This method had its first application in dentistry for the construction of false teeth, bridges, and crowns. Several decades later it was first used in making jewelry; it is now a basic process for the mass production of costume jewelry.

Other industrial applications have caused centrifugal casting to be developed as a highly controlled, almost automated technique. Precise and accurate forms in many nonferrous metals and alloys are manufactured in this way for the electronic, aeronautical, automotive, and surgical tool industries. Even before the negligible machine-finishing of the cast objects is done, tolerances of 0.002″ to 0.005″ are possible.

The equipment for such controlled work is costly and extensive. Adequate centrifugal casting equipment for the personal shop, limited as it may be when compared to industrial possibili-

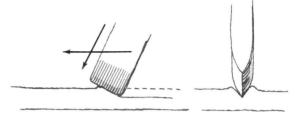

Fig. 74

ties, need not be overly expensive. However, it is true that consistent accuracy is sometimes impossible with limited equipment used under other than laboratory conditions.

Personal experience and the adaptation of much that has been learned in mass production techniques can result in very fine creative forms and this section is written to supply as much technical information as possible to this purpose.

Designing an object in wax demands an entirely different concept than working the rigid metal forms of sheet and wire. Wax is a plastic material, easily molded to the changing wish of the artist. It resists little, is receptive to every pressure or perforation, and may be used to form a replica of any other solid form in existence.

Here lies the danger. The temptation to make artificial flowers, insects, and what have you—exact in every detail— is so great that the majority of commercial jewelry designers fall victim to this ultimately boring ideal. Few artisans in history were able to literally mirror natural objects with tasteful results. Even then fantastic virtuosity of technique was necessary to compensate for the breaking of essential aesthetic laws against the slavish copying of nature.

To work in wax for jewelry one must, in a sense, adopt the sculptor's frame of reference. Since it is possible to alter mass, volume, and surface at will, these elements must be invested individually, and as a total, with the most sensitive and creative personality.

Though a nature form may be the point of departure, it should be only

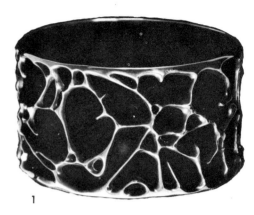

1

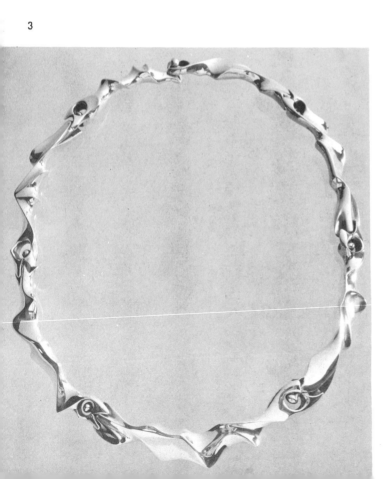

2

3

The jewelry of Ronald Hayes Pearson reflects two aspects of contemporary jewelry to perfection. First, he is a consummate craftsman. His forms, whether cast or constructed, are examples of complete control of the medium. Surface qualities enhance form and details of function have been perfectly analyzed.

Secondly, his work reflects today's concern for form; not form as it translates a recognizable object, but form which has been invented by a mind aware of the effectiveness of contour, tension, and reflection. In each piece the innate quality of precious metal is exploited to its fullest and the result is jewelry of great personality and dignity.

Mr. Pearson is one of a growing group of American artist-craftsmen who derive their livelihood through their creative work.

1. Silver bracelet; Ronald Hayes Pearson.

2. Silver and turquoise pin, 1958; Ronald Hayes Pearson.

3. Silver necklace, 1954; Ronald Hayes Pearson.

4. Silver bracelet, 1958; Ronald Hayes Pearson.

5. Necklace in 14K yellow gold, 1958; Ronald Hayes Pearson.

6. Silver and turquoise pin, 1959; Ronald Hayes Pearson.

7. Pectoral cross and chain in 14K yellow gold, commissioned for Dr. Cadigan, Episcopal Bishop of Missouri, 1959; Ronald Hayes Pearson.

Photos by Ronald Hayes Pearson

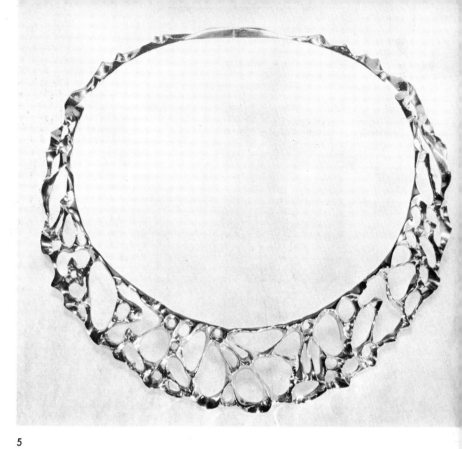

5

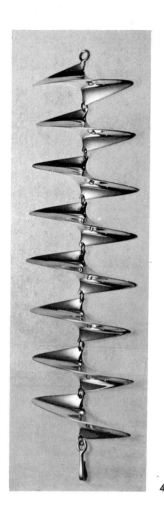

4

6

7

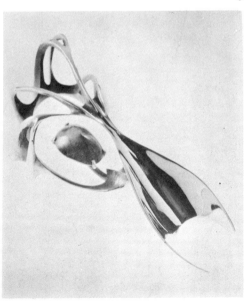

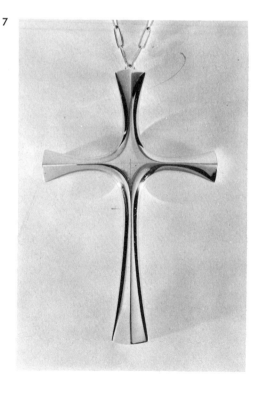

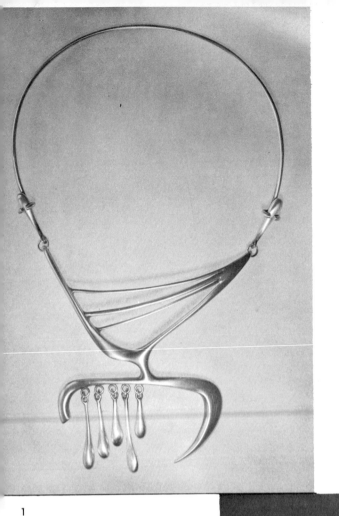

Irvin and Bonnie Burkee, working as a team, have for years maintained themselves as producing jewelers. In order to do so, they have developed limited-production techniques unique in that nothing of the priceless individuality of each work is lost through reproduction. In addition, they have created numerous thoroughly unique works incorporating many of the approaches and materials of sculpture. Not only silver and gold, but also wood, stone, ivory, and bone are used to develop jewels which miss being considered fine examples of sculpture only because of their size and function.

Both artist-craftsmen have a sound sense of form and a fine, purely personal abstract approach which reflects well the contemporary synthesis of realism and free interpretation of subject matter.

1. Collar necklace in sterling silver; Irvin Burkee.

2. Pendant in mastodon ivory and sterling silver; Irvin Burkee.

3. Collar necklace in sterling silver; Bonnie Burkee.

4. "The Circus," sterling silver pendant; Bonnie Burkee.

5. Collar necklace in sterling silver; Bonnie Burkee.

6. Collar necklace in sterling silver; Irvin Burkee.

7. Comb in sterling silver; Irvin Burkee.

Photographs by Irvin and Bonnie Burkee

1

2

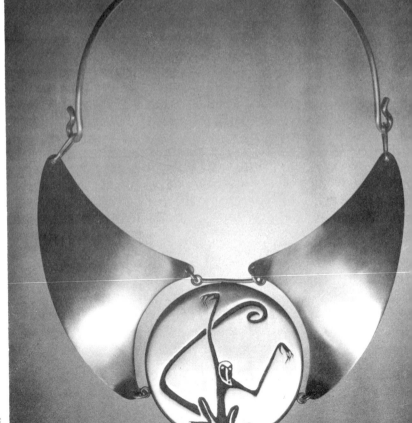

3

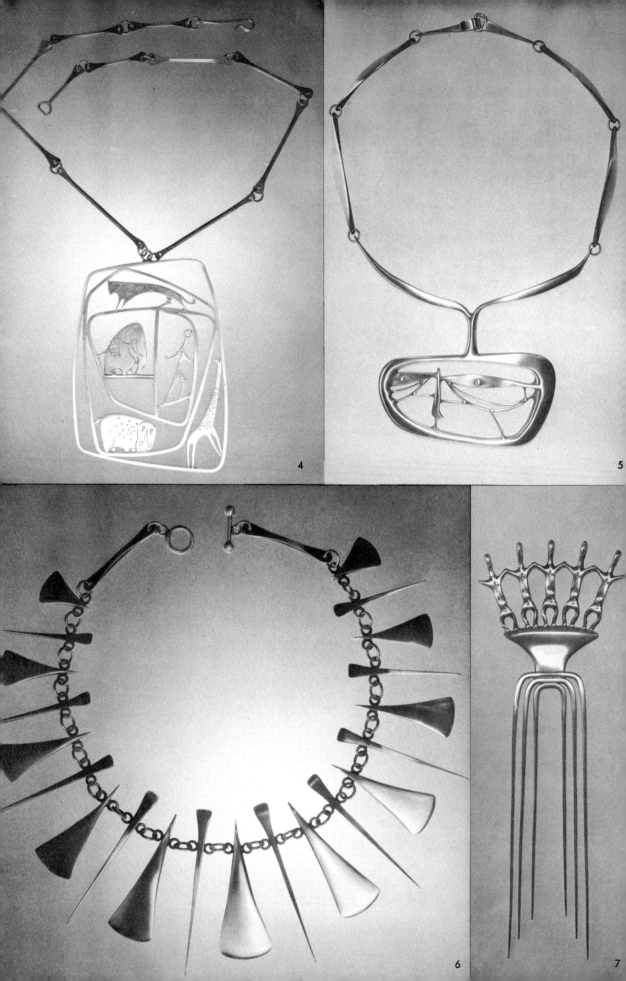

4

5

6

7

1

2

1. Cast sterling silver pendant with pearls; Earl Krentzin.
2. Cast sterling silver pendant with gem stone; Earl Krentzin.
3. Pendant of amaranth wood, opal, gold, and silver; Earl Krentzin.
4. Sterling silver pin with pearls; Earl Krentzin.
5. Sterling silver pin with moonstones; Earl Krentzin.
6. Sterling silver pin with gem stone; Earl Krentzin.
7. Cast sterling silver and gold pendant; Earl Krentzin.

Photos by Earl Krentzin

4

3

5

Though Earl Krentzin is an accomplished craftsman in many jewelry techniques, his special interest has been in casting. Much of his work, always with a highly individual humor, has religious connotations and many of his works have been designed specifically for use in religious ritual.

His free use of wood, ivory, gems, and other materials in combination with silver and gold often leads to a complexity of surfaces and textures having the precious richness of early religious metalwork.

Mr. Krentzin, in developing a personal style of expression, amply proves that individuality is possible in all art forms and that such individuality is a worthwhile and logical goal. Without such personal interpretations of design elements, work in metal too easily becomes merely skillful technique, resulting in anonymity.

7

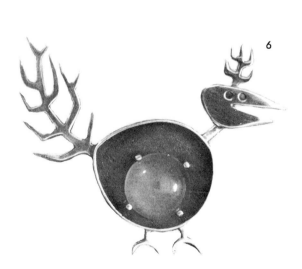

6

that and not the ultimate goal. The sophistication and the imaginative variety with which nature forms were interpreted in the lost wax casting of Pre-Columbian civilizations in the Western Hemisphere and on the West Coast of Africa are excellent examples of the healthy design possibilities.

Working in a direction devoid of conscious natural references has equal dangers. The nonobjective form has, in the great majority of cases, the stamp of anonymity. It is much more difficult to inject individual personality into a *free form* since so many combinations of tensions and planes have already become public domain. To repeat them with only minor changes is to perpetuate a cliché.

To work in the nonobjective idiom one must be well schooled in the perception of change and variety in natural forms. One must have the free will and the imagination to explore the unknown, backed by the knowledge of what has already been done.

Taken as a whole, jewelry in the *round* rather than a flat plane concept is more effective in casting. The design is limited only by the size and the volume of the casting apparatus. The steps of construction are not too flexible, so they should be followed carefully.

Materials and Equipment

WAXES

Though an adequate model may be made of beeswax, wax products have been developed that have a much greater range of plasticity, strength, and stability. Casting waxes consist of combinations of mineral and organic waxes and various gums. These may be carnauba, candelilla, beeswax and spermaceti, all organic waxes, and paraffin as a mineral wax. The gums are mastic, copal, damar, resin. The combined ingredients *must* burn away without leaving a residue.

The prepared waxes come in blocks, sheets, rods, and wires of numerous shapes. They vary in hardness and in the temperature at which they become soft or melt. Colors are used as codes for the above properties by the manufacturers, but codes differ from manufacturer to manufacturer.

Some waxes are so sticky that they cannot be carved or smoothed. These are used for combining forms or, as wires, for sprueing. Other waxes are so hard that they become adhesive only when heated to melting. These harder waxes may be softened to modeling temperature by a short immersion in water at 150° F. There is a harder wax available that may only be carved, scraped, and sanded to shape rather than formed by manipulation.

WAX HANDLING TOOLS

These may be simple and constructed for a specific purpose. Discarded dental spatulas and other shapes are often excellent tools for this purpose. A 5″ length of 14 gauge copper wire, planished and polished to a spatulate tip, may be the only tool needed. A heavy sewing needle fixed into a section of doweling or into a mechanical pencil is ideal where a small local heat application is needed. A thin knife blade, or a set of blades to fix into a common handle, may be used for carving, cutting, and smoothing. Fresh, clean emery paper of No. 1 and No. 3/0 cut may be used for final smoothing. A small piece of well-used chamois leather is also good for the above purpose.

WAX MELTING

Most craftsmen use an alcohol lamp since this burns with a clean flame which is hot enough to heat a tool tip quickly.

Candles and various gas appliances are adequate but they may incorporate so much carbon soot into the wax that the final cast metal surface may be rough or porous. Avoid a yellow reducing flame for wax working.

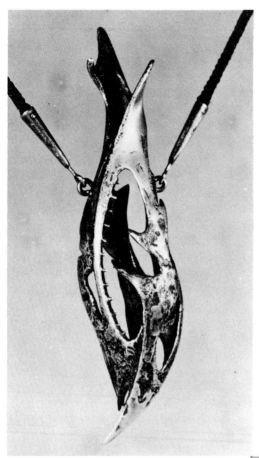

2

3

1

Christian F. Schmidt has evolved a most personal quality in his jewelry by a very logical procedure. He has examined natural organic forms and developed a great variety of interpretations from them. Each interpretation examines a fresh approach to the basic form.

In one work, surface texture, in contrast with a highly polished interior form, creates the fascinating variables found in fine art. In another, the complex relationship of solid and pierced areas develops a sculptural unity which changes with each point of view.

Casting and forming silver and gold, imbedding gems, often cut and polished to integrate with a specific metal form, Mr. Schmidt has created sculpture to be worn —beautiful in its variety and perfectly valid in its two functions.

1. "Pod Form," cast sterling silver pendant; Christian Schmidt.
   Photo by Christian F. Schmidt
2. "Pod Form," formed sterling silver and gold pendant with gem stone; Christian F. Schmidt.
   Photo by Christian Schmidt
3. "Twin Pod," sterling silver and gold casting with jade; Christian F. Schmidt.
   Photo by Robert Burningham

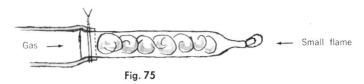

Gas →   ← Small flame

Fig. 75

Large surfaces may be made smooth by brushing them with a very small flame held at the proper distance from the work. Direct contact often destroys detail, so heat by radiation is best.

A small enough flame for the above purpose may be formed by attaching the glass tip of an eye dropper to a rubber gas hose. Packing the dropper loosely with cotton forms enough back pressure so that a very small flame may be formed. (See Fig. 75.)

SPRUE EQUIPMENT

The completed model must be mounted on a sprue former or base which eventually forms the opening through which molten metal enters the model.

Some formers are part of a base. This type of sprue former is inverted into the investment-filled casting flask. (See Fig. 76.)

A more common type of former consists simply of a small cone with a hole at its apex. It is usually filled with sticky wax or plasticene to hold the sprue pin firmly. Fig. 77B has a stepped base to accommodate casting flasks of the several diameters.

The sprue pin—the connection between the model and the sprue former —may be of 14 gauge brass wire, of finer wire, or of wax wires only. It must be firmly attached to the sprue former and to the model so that the model does not float off during the pouring of investment or during the vibration process where air bubbles are removed.

One way in which the pin may be fixed firmly into the former is to bend the end that projects out of the bottom of the former and then to press it into the wax which fills the cone. *Note:* Press the pin into the wax far enough so that the base remains level. (See Fig. 78.)

CASTING FLASKS AND THEIR PREPARATION

Heavy-duty flasks are seamless cylinders of stainless steel. Much dental casting is done in brass cylinders, but these tend to break down with heavy use and may melt if overheated.

The thinnest wall possible, with adequate strength, is most desirable, for this speeds up the wax melting process.

Casting flasks are made in many heights and diameters—from dental flasks not much larger than a thimble to industrial flasks 12″ high and 6″ to 8″ in diameter. Keep in mind that a large flask, containing much wax and a greater volume of investment, takes much longer to completely melt out than a smaller flask. The melt-out schedule must be adjusted accordingly.

If absolute accuracy is not important and if expansion or contraction have little bearing on the final outcome, a casting flask may be made of an ordinary tin can. It may be used only once—prolonged heat will have distorted it too much—but work of an awkward size or shape may be accommodated where a rigid steel flask could not be used. If a tin can flask cannot be found of just the right height, a larger one may be cut down with light curved metal shears. File or bend down dangerous burrs. It is also best to bind the can with heavy iron wire to prevent as much distortion as possible. There are times when a model, by its shape and size, must be sprued up in such a way that a cylinder

Fig. 76

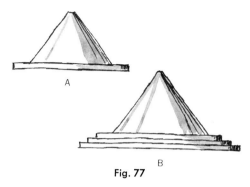

Fig. 77

will not work. A thin tin can flask may be bent into a shape to fit it. Before investing the model in the flask the flask must be lined with sheet or strip asbestos. This takes up excesses of expansion and contraction which could cause fractures in the model area if neglected.

THE INVESTMENT

The investment is a form of plaster designed to set with a hard, very smooth contact with the model. It must be porous enough to allow gases to move out ahead of the incoming molten metal. Above all, it must not break down under prolonged high temperatures or become soft enough to crack or crumble with the shock of molten metal forced in by centrifugal action.

Plaster of Paris alone will not withstand the heat requirements of an investment. The *cristobalite* investment most often used is made up of:

1. Plaster .... Used as a binder
2. Silica
   (cristobalite) High refractory properties
3. Boric acid For uniform thermal change in melting out wax. The cooling investment should contract at the same rate as the cooling metal during casting
4. Graphite .. Used to prevent oxidation

The cristobalite investment has a crushing strength of 1500 lb. per sq. in.,

strong enough to withstand casting shock if it has been mixed properly and brought to the casting temperature correctly.

VACUUMS AND VIBRATORS

Complete elimination of bubbles from the investment is necessary to prevent having the bubbles cast as metal grains on the surface of the model. Large bubbles may distort or completely destroy delicate wax areas during the wax burnout.

Industrial equipment consists of a vacuum pump and a bell jar. Often the bell-jar bed is mounted on a vibrator which dislodges bubbles from the model as the vacuum draws them to the surface of the casting investment.

If the wax model has been carefully painted with several layers of bubble-free investment, the danger of bubbles on the wax surface forming during flask investment is greatly minimized.

If a special investment vibrator is unavailable, a constant tapping with a rubber mallet on the bench top helps to move stubborn bubbles to the surface. A small vibrating sanding machine placed correctly will also work well.

THE CASTING MACHINE AND EQUIPMENT

It is best to purchase as large a machine as is practical. Though a design may be cast in sections, using a small dental machine, and then the units soldered together, it is far better to cast the total form in one operation.

Machines may be hand-driven (as the

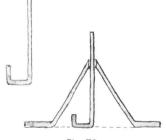

Fig. 78

earliest were), spring-activated, or motored and controlled automatically by electricity. Of the three, the spring-activated machine is best for the jeweler considering efficiency and cost. Machines may be designed for very small dental purposes or for larger general use, often with extensions for counterweights and larger crucibles available, if necessary.

The machine must be mounted carefully so that it will be perfectly level and free of vibration when in use. Bolting it to a heavy level bench at a 30″ height makes it easy to attend during casting. An unleveled machine loses much of its centrifugal force and the main bearings

crucible and the casting flask. This balance must not be ignored since uneven rotation may cause the molten metal to miss the sprue.

Crucibles of a refractory material are most often designed for use with specific machines. They are shaped to fit the crucible bed and retaining wall and, in most cases, cannot be used with other machines. When casting gold alloys, the crucibles must be kept clean by lining them with moistened strips of asbestos and relining them for each new melt. For casting silver it is best to use a heavily fluxed crucible without asbestos lining. Keep a separate crucible for gold,

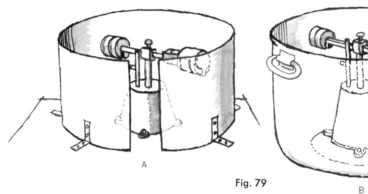

A

Fig. 79

B

will be seriously damaged. Always oil the main shaft with a light machine oil before and after each casting.

Since most machines rotate on a horizontal axis, a sheet metal barrier should be constructed to prevent accidents due to excess hot metal spinning through the area. If an 8″ slot is left open at the front of the machine, the operator may have easy access to the spring release and, at the moment of release, step to one side for protection.

Many craftsmen mount the machine in an ash bucket or garbage can bolted through to a solid surface. (See Fig. 79.)

Some machines rotate on a vertical axis and need not be surrounded as carefully by a protective wall in case of spillage.

The machine uses counterweights to balance the weight of the metal-filled

one for silver, and one for other metals. Do not allow them to become caked with heavily oxidized flux. Should this happen, the flux may be dissolved in boiling water and the crucible slowly and completely dried before casting.

BURN-OUT EQUIPMENT

This equipment is rather critical, since an excessive reducing atmosphere (excess of combustion in the presence of oxygen) causes carbon and sulfur coatings on the inner mold surfaces. These result in porous, heavily oxidized castings.

The ideal melt-out furnace would be a gas-fired, vented muffle oven. Provision should be made for the gases released by burning wax. Wax should not come in contact with heating elements if the oven is electric.

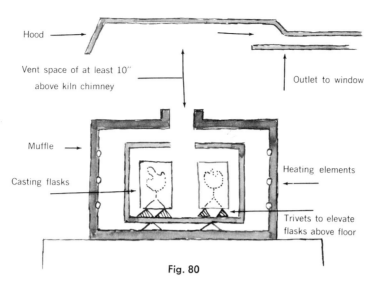

Hood

Vent space of at least 10″
above kiln chimney

Outlet to window

Muffle

Casting flasks

Heating elements

Trivets to elevate
flasks above floor

Fig. 80

An accurate pyrometer for visual heat control is an absolute necessity.

A rheostat control or timer for maintaining the correct heat level at any point in the wax burn-out is advantageous and time-saving. Without it the operator must constantly supervise the schedule, which might take as long as 8 hours.

Electric units such as enameling kilns are often used, but deposits on the heating elements caused by casting moisture and burning wax shorten the life of these elements as well as making the entire kiln interior unfit for fine enameling. If such a kiln is used, a *muffle*—a metal or fire clay chamber—should be used. In addition, some means of venting gases should be arranged. Do not connect the hood directly to the kiln chimney, for this would draw out most of the heat. There should be at least 10″ of space between the chimney and the hood. (See Fig. 80.)

Kilns or ovens without muffles or venting devices may be used if the door is used for ventilation. There is, however, a greater possibility of oxidation of the investment with such equipment.

Trivets of refractory clay or stainless steel are used to support the flasks above the kiln floor in the melt-out.

An asbestos glove and long-handled flask tongs are used in removing the hot flasks. The tongs should be designed to hold the flask securely since they must be used to place the flask into the cradle on the casting machine. (See Fig. 81.)

MELTING EQUIPMENT

Industry now uses high-frequency melting of metals for casting. The metal is in an enclosed atmosphere, is heated to flowing very rapidly, and remains completely free of oxides. This equipment is costly, so most craftsmen use torch heating for this purpose.

Whatever the heat source may be—acetylene, oxygen-gas, or air-gas—the most important factor is to use a *reducing* flame in melting the metal. An oxidizing flame will so oxidize the metal that the casting becomes rough and porous.

When using oxygen and gas or gas with compressed air, turn on the gas first, then add enough air or oxygen un-

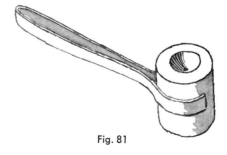

Fig. 81

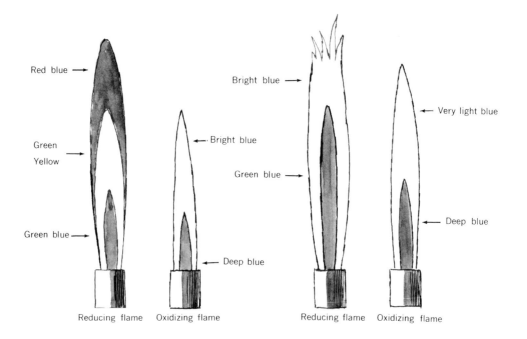

Red blue →

Bright blue →

← Very light blue

Green →
Yellow →

← Bright blue

Green blue →

← Deep blue

Green blue →

← Deep blue

Reducing flame   Oxidizing flame

Reducing flame   Oxidizing flame

Oxygen-Gas

Fig. 82

Gas-Air

til the yellow flame has just disappeared. (See Fig. 82.)

With acetylene torches, control of the air mixture is not always possible. Try for a flame that retains a tinge of yellow in the first half. An oxidizing flame is a clear blue throughout and is noisier than a reducing flame. A reducing flame will not melt metal so quickly but it is much safer.

Some authorities recommend melting the metal in a crucible in a muffled melting furnace and transferring it to the heated casting crucible just before the cast is made. This technique has advantages where large amounts of metal are cast. Some machines will take up to 100 troy ounces of silver, which would be impossible to melt in the casting crucible with a torch.

For the jeweler-craftsman, melting the metal—1 or 2 oz. at each casting—in the crucible attached to the machine is perhaps the best technique.

In addition to clean metal, various fluxes must be used. Special fluxes are designed for casting gold, silver, or other metals, and should be used as directed in each case.

Technique of Centrifugal Casting in Sequence

MAKING THE MODEL

Cold dental waxes may be carved or scraped, but when warmed by handling or hot water they may be modeled to form. Most casting waxes are not adhesive when cold so the application of small pellets to build up a form is not possible. Special waxes used by sculptors are designed for adhesiveness and should be used with a built-up technique. Dental waxes may be built up in volume by applying drops of melted wax to a surface, allowing each drop to cool before applying the next. Small grains or spheres of wax may be ap-

Fig. 83

82

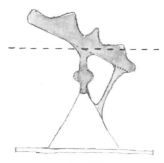

Fig. 84

plied to surfaces in this way, but, if the drop cools too much before contact, it will break off. It is best to touch the junction of sphere and form with the hot tip of a needle to form a wax weld. Dental wax wires, when warmed slightly, may be curved or rolled to any wire shape and lightly pressed to a warm surface. Again, a light touch with the needle tip insures a good contact.

The most intricate and detailed textures and linear applications may be cast. The wax may be smoothed by brushing it with a small flame, by rubbing, or by light sanding. The surface of the cast object in metal will be identical to that of the wax, so it is important to finish this as perfectly as possible. Wax may be engraved or textured with engraving tools of various shapes. Repoussé matting tools may be pressed into warm wax for texture.

Extended shapes in wax sheet or wire will cast accurately if they are strong enough to withstand the investment process and if they are sprued carefully.

SPRUEING THE MODEL

The main sprue should be attached to the heaviest part of the model. In the case of rings, however, it is best to attach the sprue to the ring shank opposite the top. (See Fig. 83.)

Use the shortest sprue possible. The sprue should be at least 14 gauge thick. A smaller sprue will *freeze* before all of the metal has entered the model. The sprue should have parallel sides but it should broaden slightly where it attaches to the model. Do not sprue to large flat areas parallel with the base—air bubbles and moisture will collect there during investment. Tilt such shapes slightly when attaching them.

There can be no areas of the model left below the point where the main sprue joins the model without the addition of a supplementary sprue. (See Fig. 84.)

A button about $\frac{3}{8}''$ in diameter should be built up at a point $\frac{1}{4}''$ below the joining of the sprue and the model. This acts as a reservoir of extra metal to be drawn upon when the interior metal contracts on cooling.

Avoid attaching the sprue to a point where molten metal will strike a flat surface in the interior of the model. The force (at least 12 lb. pressure) could damage the investment and destroy detail.

A carefully sprued model should follow the diagram shown in **Fig. 85**.

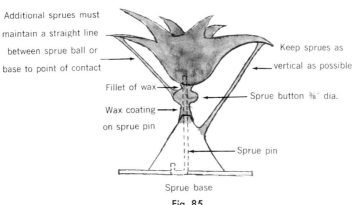

Additional sprues must maintain a straight line between sprue ball or base to point of contact

Keep sprues as vertical as possible

Fillet of wax

Sprue button ⅜" dia.

Wax coating on sprue pin

Sprue pin

Sprue base

Fig. 85

Where free wax wire forms are concentrated in a small area, extra sprueing is unnecessary. Use more casting force and *slightly* hotter metal.

The combined height of the sprue former and the sprued model is determined by the size of the casting flask.

The top of the model should be not more than ½″ from the top of the flask. This thickness allows gases to penetrate the investment and to leave the flask as the metal enters. The model should never be closer than ¼″ from the top or the sides of the flask. This would reduce the strength of the investment, causing cracks, or allowing the metal to break out completely! (See **Fig. 86**.)

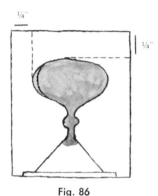

¼″   ¼″

Fig. 86

The metal sprue pin is attached to the model by heating it enough so that it penetrates about ⅛″ if possible. Reinforce the join with a fillet of melted wax.

Smooth the sprue, the sprue button, and the fillets perfectly with a small heated spatula. Rough sprueing delays and agitates the molten metal as it enters the mold.

The capacity of the casting crucible determines the ultimate mass of the wax model. Always plan to use more metal than necessary to fill the mold only, since a well-filled sprue opening assures a solid and complete casting.

It is possible to mount two or more models on one sprue or on several sprues coming from a common former. Again,

the capacity of the crucible is the determining factor.

Where many reproductions of an original model are required, a vulcanized rubber or metal mold may be made into which melted wax is injected. This is a production technique not within the province of this book. Consult the Bibliography for information sources for this process.

Removing Bubbles

The model, after being cleaned with a soft brush and cool water, must be coated with a solution which relieves the surface tension of the wax. The investment would not hold to the wax firmly if this were not done.

A mixture of green soap and hydrogen peroxide or one of the prepared *debubblizers* may be used for this purpose. The solution is applied carefully to avoid the formation of bubbles. Flowing it on from a soft brush is better than brushing it on with a scrubbing action.

The entire model may also be dipped several times into a container of debubblizing solution.

After application, gently shake or blow the wax completely dry. A very thin film of debubblizing solution remains over which the investment may be painted. Commercial solutions also control and limit excessive mold expansion during wax elimination.

Painting the Model

If no vacuum unit is used, the model must be painted carefully with investment to avoid air bubbles forming on the surface of the wax.

Mix the investment to a heavy cream consistency and vibrate the air bubbles out.

Flow the investment onto the model, making sure that complex areas are filled and free of bubbles. Build up a layer ⅛″ to ¼″ thick all around, but do not allow the investment to set completely before filling the flask. If the painting investment is too thin there will be water separation, causing rough surfaces. Investment which is too thick will not flow into small detail easily.

An alternative method, where vacuuming is possible, is to lower the model directly into the filled flask without first painting it. On very delicate work there is the danger of breaking the wax during the air elimination, so, for most purposes, painting the model beforehand is the safest procedure.

MIXING AND POURING THE INVESTMENT

The cristobalite investments may be mixed according to rigid controls to conform to desired expansion and contraction tables. This is not so important in creative casting as it may be in dental work, and many craftsmen mix investment by feel. Generally, a thinner investment is used with delicate, unpainted forms and thicker mixtures are used for simpler masses. Where a model has been prepainted, a heavy sludge consistency is best. The strongest mixture uses the least water. A thin mixture allows water to separate and accumulate on undersurfaces, causing roughness and distortion. Since dry investment changes its moisture content with age and exposure, in time accurate weighing becomes impossible.

Measure a little more dry investment than the casting flask will hold into a clean mixing bowl (a rubber bowl makes later cleaning simple). Add just enough water (about 70° to 72° F) to make a heavy sludge. Mix it thoroughly. The better investment is stirred, the stronger it will be.

Prepare the casting flask in this way:

1. Line the interior with thin asbestos sheets or strips. If moist (not soaking), it will line the flask tightly. Cut it to just overlap at the ends when in place. The asbestos lining should come to no more than $\frac{1}{4}''$ to $\frac{1}{8}''$ from the top and the bottom of the flask. This offset allows the investment to key itself to the flask itself; otherwise the whole casting might slip out of the flask during the burn-out process. The asbestos lining takes up expansion during the hardening and heating of the investment.

2. Dry the asbestos as well as possible. A soaked lining will add too much water to the investment, causing it to crumble.

3. Fasten the sprued and painted model to a clean piece of glass with a dab of sticky wax. It should not move from the center of the flask during the pouring of investment.

4. Place the flask over the model and center it. Use plasticene rolled into a thick coil to form a seal between the glass and the flask. (See Fig. 87.)

When pouring the investment tilt the flask, glass plate and all, so that the sludge slides down the side of the flask,

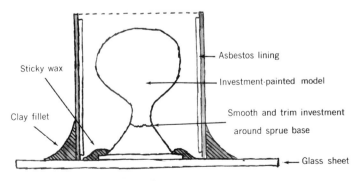

Sticky wax

Clay fillet

Asbestos lining

Investment-painted model

Smooth and trim investment around sprue base

Glass sheet

Fig. 87

covering the model from the bottom up. Agitate the flask while pouring in order to settle the sludge. Pouring sludge over the top of the model causes air pockets and might even damage a delicate form.

Vibrate the whole unit lightly (too much could separate the model from the former!) to bring up air bubbles. Then set the unit aside until the liquid disappears from the surface. Before the investment is completely set, trim off excess sludge from the top and the sides of the flask. Let the investment harden for at least 1 hour before beginning the wax elimination; 12 to 24 hours is not too long.

When the investment has set, remove the flask from the glass plate and, with a knife tip, pry the sprue former out carefully. Use narrow pliers to pull out the sprue pin if necessary.

Wax Elimination

Dental casting literature describes a short burn-out schedule of 1 to 2 hours. The flasks for most dental casting are small and contain little moisture or wax. The much larger jewelry-casting flasks require much longer burn-outs if all traces of wax are to be eliminated without undue expansion and stress. The following is the schedule for flasks 3" to 3½" high.

1. Place the flask in the furnace with the sprue opening down. The flask should be set on metal or stoneware trivets.

2. Hold the temperature under 212° F for at least 2 hours. Moisture should be eliminated gradually. Beginning immediately with high temperatures will cause steam pressure to fracture the mold or disturb the inner surfaces of the model. Most of the wax mass will have melted out during this period.

3. After 2 hours under 212° F heat, turn the flask so that the opening is up to allow the gases to be more easily driven off.

4. Advance the heat about 200° F per hour for the next 5 hours until 1200° F is reached.

5. Turn the sprue opening down again and hold the heat at 1200° F for 1 hour more.

This totals a wax burn-out time of 8 hours. Reaching 1200° F in a shorter interval usually results in the fracture of investment in large molds. This causes fins and splintering on the model surface or might destroy the interior entirely.

Since some cristobalite investments begin to break down beyond 1300° F, it is best to leave a margin of safety by going up only to 1200° F. The schedule in simple terms:

2 hours, under	212° F
End of 3rd hour	400° F
End of 4th hour	600° F
End of 5th hour	800° F
End of 6th hour	1000° F
End of 7th hour	1200° F
End of 8th hour	1200° F

The mold must be cooled to the correct casting temperature. In great part this is determined by experience, considering the factors of model size, delicacy of structure, and the metals involved.

Generally, the following flask temperatures should be reached just before the removal of the flasks from the burn-out oven:

For silver	650° F
For yellow gold	800° F
For red gold	950° F
For white gold	950° F

If the metal has been heated properly and the casting is incomplete, then the mold temperature was too low. The metal *froze* before it could reach all points of a design.

If the casting is complete but roughened or very porous, the mold temperature was probably too high—though other factors might also cause this. If the metal remains liquid too long once in the mold, it damages surfaces and contracts too greatly upon cooling.

CASTING

Before the model is invested, the correct amount of metal for the casting should have been determined. This may be done by weighing, considering that a given weight of wax is equal to 10 to 12 times as much weight in metal. Another method is to use a glass graduate to find the volume of metal needed by water displacement. Since wax floats, it must be submerged for accurate measurement. It may be forced down with a long twist of three strong wires. (See Fig. 88.)

Remember to add *at least* one fourth more metal to the crucible than would be needed to fill the model alone. A full sprue button is very important.

The metal, usually scraps, should be absolutely free of grease or oxides. Heating the metal to a bright red on charcoal and thoroughly pickling it assures a clean surface. Never use scrap containing solder of any sort. Sprue buttons from previous castings must usually be boiled in pickle before re-use.

After the metal has been weighed and *before* the burn-out begins, the machine must be balanced. Most machines have adjustable counterweights which may be locked in place after balancing.

The machine may be wound up at this point if it is spring-driven. As a rule, most machines should be wound

four times to insure solid casts, but manufacturers' recommendations vary. A release arm is put into position to hold the tension. This will be released just at the moment of complete melting of the casting metal.

When the flask is almost to the correct casting temperature, the crucible in the casting machine is preheated. Sprinkle flux into the crucible or onto the crucible liner and fuse it to the glazing point. The crucible itself should be quite hot.

Remove the flask from the oven quickly to avoid cooling and place it in the cradle on the casting machine arm. Make sure that the lip of the crucible fits accurately into the sprue opening and is pushed up to it closely. Adjustable cradles accommodate most flask shapes. This should be checked before the wax is eliminated from the flask.

With the flask in position, the heated crucible is filled with the necessary amount of metal, which should be dusted lightly with flux. Some craftsmen prefer to add the metal piece by piece as melting progresses, in the opinion that the process takes less time this way. This would depend on the type of heat used.

The metal *must not be overheated!* It should be melted enough to move easily if the arm is shaken gently, but it should

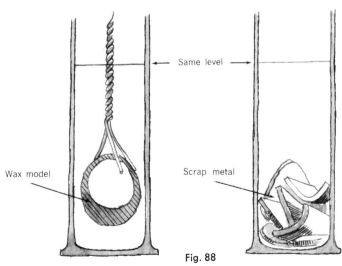

Same level

Wax model

Scrap metal

Fig. 88

never *spin* or boil. Just before release the metal should again be fluxed.

Keep the torch on the metal until after the release bar has dropped and the machine has started its spin. Never release the arm while the molten metal is moving. Allow the machine to stop its spin by itself. This insures pressure long enough for the metal to cool in place.

Remove the flask and set it aside to cool before quenching.

Silver castings should cool in the flask for 6 minutes and most golds should cool for 12 minutes.

The hot flask may now be completely immersed in cold water. The reaction causes the plaster to disintegrate, allowing the casting to be removed easily. If the investment is bone white on the ends of the flask after the wax elimination but a dirty yellow-gray or black around the casting itself, it means that the wax was not completely burned out. This causes surface porosity which might be so deep that removal is impossible.

The casting must be pickled in the standard 10% sulfuric acid solution, preferably by boiling. After rinsing and drying, the sprue or sprues may be nipped or sawed off. A successful cast will require little surface treatment before coloring and polishing.

These are sources of the common casting failures and their causes:

1. Rough surfaces and *fins:*
   a. Too much water in the investment
   b. Dirt or drops of debubblizer or water on the wax
   c. Incompletely mixed investment
   d. Wax elimination too rapid. Steam causes investment to fracture and explode
   e. Casting metal too hot
   f. Flask heated too high
   g. Vibration for bubble removal too strong or too prolonged

   h. Incoming metal striking a flat surface
   i. Model too close to edges or top of flask
2. Pitted castings:
   a. Metal not completely melted
   b. Too much flux
   c. Dirty metal
   d. Sprues too large or too small or incorrectly placed for rapid entry of metal into the mold
   e. Broken fragments of investment formed during burn-out or casting
   f. Heavily oxidized metal due to incorrect melting and fluxing
3. Bubbles of metal on the casting:
   a. Air in the investment
   b. Air trapped during the painting of the model
   c. Water or debubblizer drops on wax during painting
4. Oxidized castings (if not excessive, slight oxidation is normal on all alloys):
   a. Incomplete wax elimination
   b. Metal overheated in the casting crucible
5. Incomplete castings:
   a. Sprues too small
   b. Not enough sprues to extended areas
   c. Too much of the model below the line of contact of sprue with the model
   d. Too much investment above the top of the model
   e. Flask too cold during casting
   f. Metal not hot enough during casting
   g. Too little casting pressure
   h. Not enough metal
6. Cracks in the casting:
   a. Quenching while too hot (see cooling schedule)

Supplementary Casting Information

The craftsman may make his own alloys for casting. Pure copper is difficult to cast because it oxidizes so rapidly and heavily during melting. The addition of small amounts of silver lowers the high (1981° F) melting point and reduces oxidation somewhat without changing the color too much.

Mix the alloy before casting by melting the ingredients in a well-fluxed flat crucible. Stir the molten mass with a graphite rod and avoid overheating. Do not mix the metals in the casting crucible just prior to the casting itself!

The duplication of simple forms in wax (without undercuts) may be done by carving the form in intaglio on a smooth plaster of Paris block. After the form is well refined, the plaster is soaked thoroughly in water and melted wax is poured into the carving.

Remove the wax as soon as it is completely cool by gently prying it out. Sprue mounting proceeds as for any wax model.

Rings may be cast in this way:

1. Use a dowel 4″ long that is the same diameter as the knuckle of the ring finger.
2. Wrap the dowel *smoothly* with *one* layer of thin aluminum foil. This will prevent the melted wax from sticking to the dowel. Tape or wire the foil at both ends to hold it on the dowel firmly.
3. Cut a strip of thin sheet wax as wide as the band of the ring and just long enough to join. Fuse the ends together carefully with a hot spatula.
4. Drip wax onto the wax band until enough bulk is formed for modeling or carving.
5. Finish the model as completely as possible. Remember that the final wax model will be reproduced in metal exactly.

Setting a stone in a paved setting is quite simple.

1. Have the bulk of the ring or other jewelry form constructed.
2. Cut a flat bed for the stone. Leave a hole through the model if the stone is translucent or transparent.

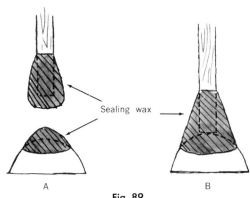

Sealing wax

A   B

Fig. 89

3. Fix a small stick to the top of the stone with sealing wax or dopping cement. First heat the stone carefully, drop on a little cement, reheat and apply the stick. Allow the cement to cool. (See Fig. 89.)
4. Place the stone with the attached stick in position in the bed in the wax and build up the wax around it. Do not hide more of the stone than necessary, but allow enough wax around the base of the stone to key it in. (See Fig. 90.)
5. Remove the stone from the finished model by first softening the wax in *warm* water. Next, remove the stone carefully with a straight, parallel, upward pull.
6. Remove the ring from the dowel and finish the inside where necessary. Sprue it and cast.
7. With the holding stick still in position, replace the stone in the finished ring and planish the metal tightly around it with chasing tools. Stone or burnish the tool marks away. Sometimes a casting might have contracted so that the stone no longer fits. Engraving tools or scrapers are used to enlarge the hole in this case.

Stones may be cast in position when incorporated in the wax model.

Soft stones, such as turquoise or malachite, and those with many fractures such as opal or labradorite, are not suitable since they would change color or shatter.

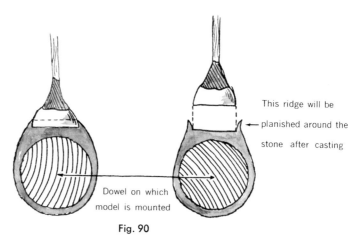

This ridge will be planished around the stone after casting

Dowel on which model is mounted

Fig. 90

Tough stones of the hardness of 8 on the Mohs scale are suitable if handled carefully. The important factors are that the flask containing a stone should be heated as slowly as possible during wax elimination. At the risk of a rough casting the metal should be cast while the flask temperature is high—at least 1000° F. The metal should be just hot enough to be fluid.

The cooling period is of great importance. The kiln or oven used in the burn-out should be brought up to 1500° F in time to replace the metal-filled flask immediately after casting. Allow the flask to return to room temperature in the closed kiln. Never quench the hot mold with a stone in place.

The accidentals formed by dropping melted wax into cold, warm, or hot water may often result in richly organic cast forms which may be used separately or combined by soldering.

FUSING AND MELTING
Fusing

When done with sensitivity and discrimination, a technique of combining sheet or wire forms by heat without soldering can result in controlled forms which turn an accident to a purpose.

If scraps of sheet metal or wire are combined in a purely emotional manner, letting them fall where they may, the result might be intriguing but it is seldom a serious statement of the artist's intent. In this way the forms are limited to the occasionally exciting accidents of natural objects—and become anonymous.

With a form concept in mind the craftsman may start with a shape or a combination of shapes, perhaps sawed out quite carefully, which establish a framework within which the technique is contained.

With careful heat control it is possible to fuse two surfaces together by causing a surface metal flow which bonds the forms. The surface, upon cooling, has the characteristic *orange peel* roughness of overheated metal, but it may be filed, stoned, and buffed where highlights are needed.

The surface of a sheet form may also be heated until it is semifluid, at which point wire, also heated almost to melting, may be added. The wire, when hot enough, becomes very flexible and may be built up on itself to form high areas.

Where a design requires controlled placement of wire, pieces may be precut and formed and then fused into place in the same way that sheets of metal are combined. Balls of shot, perhaps of different metals, may also be attached in this manner. Silver filings may be placed or sifted onto sheet at the right moment which, when fused, form a surface texture contrasting with the roughened flat areas. Filings of metals with a higher melting point than the base form may also be incorporated.

Melting

Another accidental technique consists of dropping fully molten metal into water of various temperatures and depths. The height from which the metal is dropped also controls the resulting forms to a degree.

A depression carved into a charcoal block, with a pouring groove to the edge added, makes an adequate melting crucible.

Again, the successful use of these accidentals depends on the selectivity of the artist. It is very easy to be satisfied with superficial results in the name of "happy accident" or "freely evolved form."

4 · DECORATIVE SURFACE TECHNIQUES

Though form and shape alone may result in handsome jewelry, the artist may wish to fulfill the decorative function of jewelry through surface enrichment. Often the innately precious quality of a jewel is best stated by the delicacy and variety of the surface treatment.

As in all other visual arts, the surface and the body of a jewel must relate to each other completely. A virtuoso treatment of the planes of an unimaginative basic form can only result in an unfortunate veneer—never a total statement.

The great danger in surface decoration lies in overdoing it. Understatement is always more successful since the viewer must then focus his own aesthetic awareness on the object to complete the total unity.

Perhaps the earliest forms of surface decoration consisted of embossing thinly hammered sheets of metal into figurative or geometric lines and shapes. Later, through the art of soldering, surfaces were covered by wire or metal fragments to create the rich play of light and dark so advantageous to jewelry. The casting techniques of the time translated the soldered wire into wax coils which were handled in the same decorative manner.

When the addition of fused enamel became possible, even more surface enrichments were added, with not only texture but also color playing a role.

Engraving, inlaying, lamination, granulation, and many combinations of techniques—all combined in time to give the jeweler a great range of expressive possibilities. Many of these techniques were popular for relatively short periods in history, but as potential means for contemporary use all of them should be re-examined in the light of fresh attitudes toward decoration. Some of these techniques became so important to a culture that great skills in the handling of tools and materials resulted. Volumes have been written and long apprenticeships were often necessary for a jeweler to grasp the total possibilities. Other techniques have—in all but their museum identities—been lost to us in good part. With the help of the historian, the chemist, and the metallurgist, much has been reconstructed, but even more must be relearned by individual experimentation.

Materials developed in our time, such as the many synthetic plastics, have seen only the beginnings of use as expressive media. Here, knowledge of past uses is nonexistent and the artist-craftsman must use his own training in finding

Fig. 91

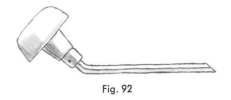

Fig. 92

adaptations and new directions of expression.

In this chapter, then, are found techniques described as they were once used and as they may be used today.

ENGRAVING

Engraving is one of those techniques which could be considered to be an art of its own. To engrave skillfully one must spend considerable time in practice alone. The occasional use of engraving on a piece of jewelry cannot teach the craftsman more than the fumbling rudiments.

There are excellent manuals available which explain in detail how to engrave the most intricate lines and how to properly shape and maintain the cutting qualities of engraving tools. Only constant practice on a variety of metals and surfaces can translate the written explanation into usable knowledge.

Tools and Their Maintenance

The tools of engraving are basically steel rods of various sections with wooden handles attached. They may be straight, for work on convex surfaces (Fig. 91), or angled for engraving flat or concave surfaces (Fig. 92). The wooden handles may be spherical or hemispherical in shape, depending on the intended use of the tool.

The cutting tool itself, called a graver or burin, might be filed, ground, and stoned to a great variety of shapes. (See Figs. 93 and 94.)

The degree of angle on the cutting tip of the tool is very important. The best angle is one of 45°. Less than 45° causes a deep bite when cutting but preserves the point quite well. More than 45° makes a shallow and easier cut but the point breaks quite often. (See Fig. 95.)

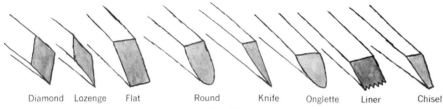

| Diamond | Lozenge | Flat | Round | Knife | Onglette | Liner | Chisel |

Fig. 93

Most tools must be shaped and sharpened after purchase. If a tool must be bent—for engraving flat or concave surfaces—it may be heated to cherry red and bent against an asbestos pad while being heated. Avoid twisting the side or center lines; keep the sides completely parallel. Quench the graver completely in water as it comes from the tap. The face and the cutting angle may now be filed or ground by using an India stone and oil. Polish one third of the graver from the cutting end with emery paper (from No. 1 down to No. 4/0). Heat this one third with a blue, soot-free flame until a straw-yellow color develops on the

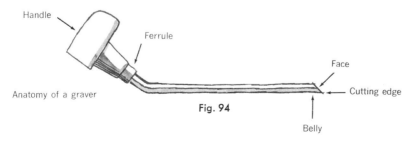

Handle

Ferrule

Face

Anatomy of a graver

Cutting edge

Fig. 94

Belly

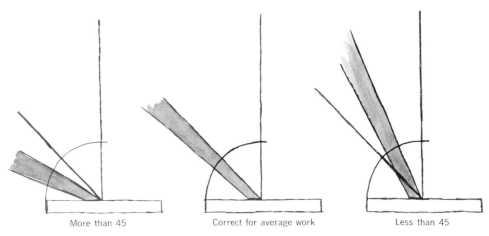

More than 45   Correct for average work   Less than 45

Fig. 95

metal and travels to the end. Quench the steel quickly in water.

The tool may now be sharpened by careful rubbing on an oiled *Arkansas stone,* with an oil such as SAE 10. Use enough oil so that metal particles do not embed themselves in the stone. The final polish to the face and the cutting edge is given by a very controlled rubbing on fine emery paper (4/0) placed on a sheet of glass. Do not round the corners—this will dull the tool.

There are grinding devices available, some of which are calibrated to allow for the setting of desired angles, which may be used to hold the graver for all but the final emery polishing.

Gravers may be *heeled* so that they turn well to the right or to the left. Heeling is done by grinding and stoning an angle to the face top and the cutting edge so that the tool naturally makes a curve to the right or the left. The many heeling angles necessary to script letter engraving may be studied in good engravers' manuals.

Design Transfer

A scribe may be used to *lightly* incise guide lines which may have been freely drawn on the metal with ink or pencil.

Chinese white or white tempera paint may be painted over the surface to take the marks of carbon paper. If tempera is used, a scribe should incise the lines since the pigment would chip off dur-

ing engraving. Chinese white, rubbed onto the metal in paste form with the finger tip, may be engraved directly.

Tool Handling

The object to be engraved should be firmly fixed to a surface that will allow the object to be turned easily. An engraver's ball or block, set in a leather ring, is ideal for small pieces, but it is expensive.

The object may be heated slightly and sealed to a wood block onto which a 1/4" thickness of sealing wax has been applied. The wax is allowed to cool and, after being engraved, the work may be pried off gently or reheated and lifted off. The small amount of sealing wax adhering to the metal may be dissolved with benzene. (See Fig. 96.)

A pitch bowl or tray may also be used and the metal applied and removed as described in the section on repoussé techniques.

A cork- or lead-lined vise or a ring clamp may be used for objects in the round.

During engraving itself, one hand steadies and turns the working surface while the other manipulates the graver. Many engravers hold the tool virtually in one place while moving the object into new positions. This is especially true for cutting curved lines.

The tool hand holds the graver so that the handle is comfortably cradled in the palm. The thumb acts as a brake while the fingers control the angle and the position of the tool. Sometimes the thumb of the holding hand is pressed against the tool thumb as a pressure control, but a slip of the tool could cause a nasty gouge.

The entire action—the press forward, the downward pressure to bite the tool into the metal, and the turning action of the tool and the metal—should be done slowly and with complete control. Since a slip with a graver actually removes some of the metal, it is quite difficult to obliterate mistakes by burnishing.

Each cut should end with a slight release of forward pressure and a slight depression of the tool. In this way the tool rides up again and leaves no burr. If a burr remains it may be cut off with a scraper held very flat to the metal to avoid shaving away the surface itself, or a cut from the opposite direction—and joining the first—may be used.

Lubricate the graver often by touching the tip to a cotton swab soaked with a very light oil. Wintergreen oil has long been used by engravers for this purpose.

If a tool leaves a *heel mark*—a scratch behind the start of a cut—it is not heeled enough for the surface. This may also be remedied by bending the entire tool at a point about 1″ from the ferrule on a 3½″ tool.

Prying up the point too quickly from a deep cut causes the point to break. This also occurs when the face angle is too flat, making the point long and narrow. Improperly tempered gravers, being too hard and brittle, also tend to lose the point easily.

A broken or dulled point must be sharpened with stones and emery paper again; make sure that the original angle (if correct in the first place) is maintained.

The sharpness of a point may be tested by pushing it into the surface of the thumb nail lightly at a flat angle. If it is deflected it is not sharp enough for clean engraving.

Other Uses of the Graver in Jewelry

Aside from the commonplace inscription engraving on jewelry, the graver is useful in a number of ways:

To create a line in surface design.

To develop a surface texture by cross-hatching or other repeated cuts.

To cut depressions and lines for inlaying niello, enamel, or other metals.

For removing solder from between fine wire areas.

For pushing up burrs in *bead* settings of small stones and for *bright cutting* planes around faceted stones.

For leveling top edges of bezels after they have been burnished around stones.

For enlarging bearings (depressions) into which stones will be set, as in cast jewelry.

ETCHING

When man first noticed the action of acids on metal is not known, but it is known that at an early date in their culture pre-Columbian Indians used oxalic acid obtained from plants to dissolve copper on the surfaces of copper-gold-silver alloys to achieve a pure gold surface.

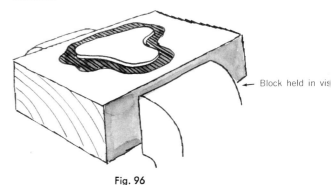

← Block held in vise

Fig. 96

In more modern times, the technology of chemical manufacture has resulted in the development of many new acids that may be used on metals in a controlled fashion.

The process is simple. The metal to be etched is cleaned as thoroughly as possible of all grease or oxides. The areas to remain as raised surfaces are covered with an acid-resistant substance. The entire piece is immersed in an appropriate acid solution (depending on the metal) and allowed to remain for as long as necessary for the acid to dissolve the exposed metal to a desired depth. The result should be—after the resist is removed—a surface composed of raised and lowered areas which carry out the meaning of the design.

Acid Solutions

The solutions—called mordants—used for ferrous and nonferrous metals are:

1. For gold of low karat (18K or lower):

Hydrochloric acid . .	8 parts
Nitric acid	4 parts
Iron perchloride . . .	1 part
Water	40 to
	50 parts

2. For silver:

Nitric acid	100 cc.
Water	300 to
	400 cc.
or	
Nitric acid	1¼ pts.
Water	1 pt.
Isopropyl alcohol . .	¼ pt.

Agitate the work constantly. The temperature should be over 75° F.

3. For copper:

Nitric acid	1 part
Water	1 part
or*	
Potassium chloride .	2 parts
Hydrochloric acid . .	10 parts
Water	90 parts

This forms a slow even etch which does not underbite as readily.

4. For brass:

Nitric acid	1 part
Water	1 part
or	
Sulfuric acid	10 oz.
Nitric acid	2 oz.
Hydrochloric acid . .	2–3 drops
Sodium dichromate .	⅛ to
	¼ tsp.
Water	1 gal.

5. For aluminum:

Ammonia	¾ oz.
Copper sulfate	2½ Gm.
75% Phosphoric acid	1–2 drops
Sodium hydroxide	7 oz.
Water	1 gal.

Etch for 3 minutes at 145° F. Rinse the object well and immerse immediately in a 40% nitric acid pickle (40 parts acid + 60 parts water). Rinse again and dry.

6. For steel or iron:

Hydrochloric acid . .	2 parts
Water	1 part
or	
Nitric acid	1 part
Water	10 parts
75% Phosphoric acid	1–2 drops
	per gal. of
	the above
	mixture
or	
Nitric acid (commercial)	50 oz.
Isopropyl alcohol . .	8 oz.
Water	1 gal.

Use at room temperature if over 70° F.

Resists

The resists are generally composed of wax, vegetable gums, or asphaltum. Some formulas are:

1. Beeswax. Immerse the entire object, once it is chemically clean, into molten wax. Warm the metal before immersion to avoid heavy deposits of wax. After etching, beeswax may be removed by heating the object to remelt the wax

and then wiping it off. It may also be dissolved off with benzene or acetone.

2. Wax pencils may be used for laying out delicate lines to remain unetched.

3. Thin shellac may be painted or dipped on.

4. Sealing wax (basically shellac) may be dissolved in alcohol by saturating 1 oz. of alcohol with shavings of sealing wax. Stir the resultant thick liquid thoroughly and brush it on. Add alcohol if it becomes too thick. Use alcohol as a solvent after etching. This is a good resist for the edges and the bottom of an object, but it tends to chip when it is scratched in designing.

5. Gum guaiacum dissolved in alcohol in the same manner described in formula 4 makes a good flexible resist. Use a strong solution of sodium hydroxide as a solvent.

6. A varnish composed of:

Liquid asphaltum ...... 16 oz.
Benzene .... 3 oz.
Turpentine .... 4 oz.

has strong acid-resistant qualities but should be dried very thoroughly with a little warmth before immersion in the acid bath. If it is still wet it lifts from the metal, causing *underbiting*. Remove this resist by boiling the object in a strong lye solution.

7. A *ball ground* resist, often used by copper plate etchers, is composed of:

Beeswax .... 2 parts
Asphaltum .... 2 parts
Burgundy pitch .... 1 part

The ingredients are melted together and, when cool enough, rolled into balls about 2″ in diameter. They are then enclosed in a bag made of a square of finely woven cloth. The ends are brought up and tied or wired together tightly to make a short handle. While the ball ground is still soft it should be pressed on a flat, heated surface so that the ground penetrates the cloth. (See Fig. 97.)

Fig. 97

The well-cleaned and polished metal is then heated to a temperature that will melt the ground but not cause it to burn. The ball is rubbed over this heated surface to deposit a fairly even layer of ground. Additional heat should spread the ground on the metal in an even, translucent layer or it may be rolled smooth with a soft brayer (roller).

After cooling, the object should have the back and the edges painted with a liquid resist. The ground may be darkened by carefully smoking it over a small turpentine flame. Do not scorch the ground! Smoking makes the line, scratched by a needle or knife down to the bare metal, more visible.

Some etchers build a small wall of wax around the area to be etched. This contains the acid so that edges and other areas need not be protected.

The process of etching varies with the mordant and with the metal. In general, the object is left in the bath long enough so that a line has been etched deeply enough to be felt with a steel point. Remove the object from the acid before testing.

Bubbles are formed by the action of the mordant and should be brushed off as they form, for if they are allowed to remain they cause uneven biting. A feather may be used to brush these bubbles away.

Watch constantly for particles of ground that may have broken away. If this occurs it is best to remove the work, remove the ground, and reapply it more carefully. Sometimes a violent biting, indicated by excessive foaming, causes the ground to break away. Dilute the formula a little at a time and try etching again until bubbles form slowly and evenly.

A pyrex baking dish makes an ideal etching bath. Work may be lowered into it on a cradle made of a loop of cotton string. Do not use wool or synthetics since many of these fibers dissolve in nitric or other acids. (See Fig. 98.)

Jewelry Uses of Etching

1. As a decorative surface defining line and form.

2. To create depressions into which enamel, niello, or other metals may be inlaid.

3. To texture large areas of metal. A long etch in a fairly strong mordant results in a finely pocked surface which holds coloring solutions well.

4. To texture or model the metal surface for *basse-taille* enameling techniques.

FILIGREE

The filigree process has had almost world-wide application. The Near East has for centuries been the focus of skilled workmanship in this delicate art. Filigree may consist of wire soldered to wire with no background support, or it may be wire soldered to a larger metal surface. Traditionally, flat wire has been used more often than round or square

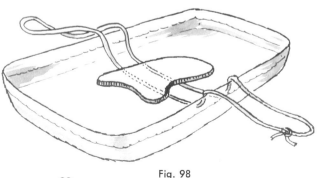

Fig. 98

98

wire. Since it is flat it can be bent to shape easily with fingers and tweezers. The wire is usually very delicate; perhaps 22 gauge by 28 gauge. When fine wire is to be applied to a contoured surface, a round wire of the desired gauge is easier to use since it bends to the surface more readily in all directions.

Fine silver and high karat golds are most often used for filigree because of their malleability. Sterling silver or low karat golds must be annealed thoroughly if they are to be flexible enough.

It is possible to purchase well-annealed wire of virtually any gauge from the refinery, but often a need for a specific gauge arises that cannot be met quickly in this way. Drawing the wire to the desired gauge through a drawplate is a simple matter. The tools needed for wire drawing are:

1. *A drawplate*. This is usually a bar of hardened steel about ¼″ thick which has been perforated with tapered holes that increase in size from one gauge to the next. The holes may be round, square, oblong, triangular, half round, or elliptical. Some drawplates have a good range of holes of several shapes.

The drawplates used in the large manufacture of wire have the holes drilled through a hard mineral such as corundum (ruby and sapphire).

2. *Draw tongs*. These are specially designed pliers that have an easily gripped handle and serrated jaws for secure holding of the wire as it is pulled through the dies in the drawplate. Most draw tongs have a blunt, squared-off nose to enable the wire drawer to grip a maximum amount of the wire projecting through the die.

3. *Beeswax*. This is rubbed into the holes of the plate or onto the wire to lubricate the wire as it is pulled through the die. The heat of friction and compression melts the wax into a covering film. To draw wire without a lubricant is both more difficult and also harmful to the drawplate.

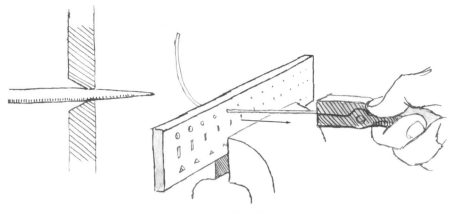

Fig. 99

Preparing and Drawing the Wire

The wire must be pointed by grinding or filing so that enough projects through the die for the tongs to grasp firmly. The taper of the point should start 1″ back from the wire end.

The drawplate should be placed into a smooth-jawed vise horizontally. If inserted vertically a hard pull on the wire could cause an expensive drawplate to break in half. The vise should be attached to a heavy bench—since much strength is sometimes necessary to draw heavy wire, a light bench could be pulled around the workshop quite easily during the process.

Place the pointed and lubricated wire end through the first hole into which it does not fit easily. Grasping it firmly with the tongs, draw it through with a steady pull, keeping the wire perpendicular to the drawplate at all times. (See Fig. 99.)

Move to the next hole in line. Do not skip a hole since this will cause the wire to break if it can be pulled at all!

Since the wire is compressed and thus elongated it soon becomes brittle.

To soften wire by annealing, it should be wound compactly into a coil. A length of silver wire is wrapped around the coil to prevent expansion and to keep the wire solidly tight. Loose loops are easily overheated during annealing. If iron binding wire is used, it must be removed before pickling, so a length of sterling wire, which may be used again and again, saves time. (See Fig. 100.)

The coil may be painted with a temperature-indicating flux, which becomes water-clear at 1100° F. Placing the coil on a thin nickel-chromium sheet or wire grid allows the torch to heat all sections of the coil evenly.

If a kiln with an accurate pyrometer is available, the coil may be placed on a trivet and brought up to annealing temperature very safely. This has the advantage also of heating even the interior of the coil thoroughly and with a minimum of oxidation.

As soon as the correct temperature has been reached, the coil must be quickly quenched in cold water or pickle.

After rinsing the coil, dry it thoroughly to prevent rust in the drawplate dies if additional drawing is to take place.

The wire is gently unwound and straightened and drawn through the dies until it again becomes too stiff and springy.

Two annealings usually suffice when drawing wire from gauge 10 to gauge 28 B and S. The first annealing takes place when the wire has become tough and the second when the drawing is complete. Unannealed wire is almost impossible to bend accurately.

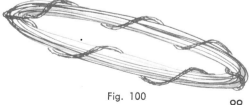

Fig. 100

Cut number 1

Cut number 2

Fig. 101

Soldering Filigree

Very small pieces of solder and a small accurate flame are necessary when soldering together the delicate wire forms of filigree. Cut sheet solder, usually *Easy*, into very small *paillons*. Pieces $\frac{1}{64}''$ square are not too small. Be sure that the thickness of the sheet solder is less than that of the wire. Solder may be rolled thin through a rolling mill or planished with a smooth-faced hammer on a steel block. The solder should be clean to prevent oxidation.

There are many ways in which filigree designs may be set up for soldering. Most craftsmen lay out the pieces on a smooth charcoal surface and heat from the top. Others glue the pieces to a thin sheet of annealed steel or nickel-chromium with a paste made up of a little gum tragacanth and flux (liquid flux, since it does not foam as readily, is better than paste flux for this purpose). The sheet is placed on a tripod and the work is heated from below. The advantage here is that small pieces are not easily blown out of place by the flame.

A third, more complicated technique, consists of:

1. Pressing the wire shapes about one fourth of their diameter into a thin, flat sheet of beeswax.

2. A box wall is built around the impressed forms with thin cardboard and the enclosure is filled with a heat-resistant plaster such as is used in casting.

3. When the plaster is hard, the wax sheet is pulled away cleanly, exposing the wire again. After the plaster is completely dry—*important!*—the joins are cleaned with a thin knife blade and soldered as usual.

The plaster will dissolve in cold water if quenched immediately after soldering, or it may be soaked away in a very dilute sulfuric acid pickle. A light brushing will remove plaster from hard-to-reach places. The work may be cleaned by boiling it in pickle.

Remember that solder will flow to the hottest point, especially when wire is being joined. Heat the *entire* piece equally. A small-pointed steel scribe may occasionally be used to move solder into a join just as it flows.

Previously soldered areas may be protected by a thin paste of yellow ocher when additional soldering is necessary.

INLAYING

The contrast of color or texture may play an important design role in a piece of jewelry. Dark metals may be combined with light-colored metals with dramatic results. Wood, ivory, and plastics may create not only a color contrast but also a surface change when inlaid in metal or into each other.

Inlaying Metals

Soft metals have been inlaid into depressions in bronze, iron, and steel for centuries in many parts of Europe and Asia. Much of the impressive armor of the Middle Ages and the Renaissance was decorated in this way. Brass and copper inlay work is still practiced in

Fig. 102

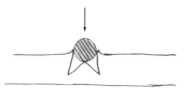 

Fig. 103

the Moslem countries and in India in much the same manner as it has been practiced for thousands of years.

There are many ways of laying one metal into another. New variations may be developed occasionally with advances in metallurgy. Perhaps the earliest technique consisted of carving or engraving lines and depressions into which another metal was hammered. This method is still worthy of attention today.

For inlaying wires into flat or curved surfaces, engrave a line with two cuts of the graver. If a burr results, so much the better. The cuts should be narrower at the surface than at the bottom. (See Fig. 101.)

With a narrow-pointed chisel or graver, the bottom of the engraved line should be *plinked* with small pointed projections every ½″ or less. The softer metal of the wire will be pierced by these sharp points, keying it in place. Fine silver, gold, or other soft annealed wires are then hammered into the groove with a small, smooth-faced hammer. The hammer face should be slightly convex so that a tilted blow does not mar the metal. Punches of the right shape may also be used for this. The wire should be a little thicker than the depth of the groove so that it will move into the undercuts. (See Fig. 103.)

Large flat inlays are set in a similar manner. The depression may be etched to the desired depth or carved out with a chisel or graver. The sides should be undercut with a graver in any case. The small points should be raised about ¼″ apart and pointing in many directions.

By the laying of a small polished steel block over the entire inlay it may be planished down evenly without the danger of hammer marks.

A simpler technique consists of making the base form of two sheets, one solid as a backing, and one with the depressions cut out with a saw. The two sheets are soldered together carefully; make certain that the solder runs to all edges. The bottom is plinked with a graver and the inlay hammered in.

Shot may be used as the inlay material by drilling and filing a hole into which it fits tightly. A few taps of the hammer spread the shot into the undercuts of the hole. (See Fig. 104.)

A variation consists of filing the inlay piece so accurately that it may be soldered into place without hammering.

When inlays are to be hammered, they should fit perfectly but should be one or two gauges *thicker* than the depth of the depression. This allows the metal to spread into undercuts. If differences in surface remain after hammering, they may be filed and stoned away. If it is impractical to use a metal of thicker gauge for the inlay, the inlay should be domed slightly and filed to fit exactly before hammering.

Another technique, as venerable as it is dangerous, combines finely powdered gold or silver with an equal amount of mercury. The ingredients are worked in a mortar with a pestle until they are of a heavy pastelike consistency. This amalgam is burnished into the prepared depressions and kept at a temperature of about 150° F for two or three days. This must be done with great care and with

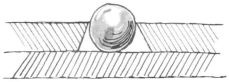

Fig. 104

good ventilation since mercury vapor is highly poisonous. After a few days most of the mercury has evaporated, but the entire piece should be heated to 1100° F to 1200° F to remove the last traces. The inlay metal may now be burnished again and the whole form polished.

Dental silver amalgams as used in filling cavities may be used for this purpose, but such metal is never so white as sterling or fine silver.

A very simple inlaying technique uses solder as the inlay. The foundation metal must have a high enough melting point so that the solder can become completely fluid before the foundation metal nears overheating. Use a liquid or powdered flux mixed with filed solder and use a larger amount than the depression will hold to compensate for the decrease in volume when the solder melts.

The possibility of using patinas and coloring agents to enrich an inlaid design should not be neglected. The Japanese *tsuba* or swordguard decorations demonstrate very dramatically the richness of gold on a heavily blackened background or the subtlety of silver surrounded by a rich red-brown copper patina.

Inlaying Nonmetallic Materials

Wood, ivory, mother-of-pearl, and other organic materials may be inlaid in metal effectively.

The inlay procedure using two sheets as the base is most effective for this purpose. Pieces of fairly soft materials (such as wood, ivory, or bone) may be forced into the inlay depressions with light blows of the hammer or punch. The use of a good mastic cement, as it is used in fastening formica to wood, is necessary for inlaying fragile shell and thin sections of stones or gems. In the latter case, the material should be cut and filed to fit perfectly or, where this is impossible, the depression made to accommodate the material. Solid plastics may be used as wood is used and liquid plastics, colored or clear, may be poured into lines and depressions. The champlevé enameling technique is also an inlay process.

A method by which fragments of broken gem stones of certain types are inlaid in metal depressions has recently become widely used in Mexican jewelry, though it has been an important technique in North Africa and India for centuries. A depression is created by the soldering of a perforated thick gauge of metal to a backing sheet or by construction of an enclosure of flat wire. (See Fig. 105.)

Fig. 105

Lumps of raw gem material (such as turquoise, malachite, or chalcite) are broken into small—$\frac{1}{8}''$ diameter or less —fragments. These are packed into the depression in such a way that a large number of flat surfaces face upward. Sealing wax—which comes in several colors such as red-orange, black, green, etc. —is crushed to small grains and sifted into and over the imbedded gem material. (Be sure to use enough sealing wax to allow for shrinkage of volume when the wax melts.) The work is gently heated from below until the wax melts, thus cementing the gem particles securely. Filing, stoning, and polishing develop a handsome, level surface of contrasting texture and color. Liquid thermosetting plastics may be used instead of sealing wax in this process.

LAMINATION

This form of surface enrichment, not used often today, exploits the *planned accidental.*

The process is simple but time-consuming.

1. Four or five flat sheets (1" x 3" x 20 gauge is a good size) of different metals and alloys are combined by soldering. Copper, bronze, silver, and alloys of gold and silver or gold and copper form a good color range.

2. Dark colors are alternated with light colors. For instance, the bottom layer may be of silver, the next of copper, the next of silver, the next of bronze, the last of silver. Thinner gauges of one or more of the metals may be used if more variety is needed, but the soldered bar should not be so thick as to make the rolling process difficult.

3. Starting with the first two layers, the soldering may be done in two ways:

(a) Small *paillons* of *medium* solder are placed $\frac{1}{8}$" apart around the edge of the cleaned and fluxed bottom piece. Additional solder is placed $\frac{1}{4}$" apart in the interior. The second sheet, also fluxed, is laid onto the first and they are soldered together with a large, soft flame; then they are pickled, rinsed, and dried. The solder is now applied to the second piece and the third soldered to the first two, and so on. Where thinner than 20 gauge pieces are used, three or more layers may be soldered at one time.

During each soldering, be very sure that complete contact between the sheets has been made. When the solder is completely molten, raised areas may be pressed down carefully with tweezers if necessary.

Be very careful when soldering brass and some of the bronzes. They may suddenly collapse into the silver or the copper with little warning.

(b) A *eutectic* solder (a combination of a flux and powdered solder) may be painted thinly over the contacting surfaces and thus soldered together. Use a eutectic solder of at least 1350° F flow point.

This method usually results in evenly distributed solder throughout the join. This is important for some of the later processes in lamination.

4. The bar is now rolled through a rolling mill until it becomes quite stiff and resists the thinning process. Roll it through in as many directions as possible to keep the stresses even throughout.

5. Once the bar becomes hard to roll, it must be carefully annealed.

6. After annealing it should be pickled, well rinsed, and completely dried. If water or pickle comes in contact with the rollers of the rolling mill, serious damage to the polished surfaces will result. Keep a light film of oil on the rollers at all times.

7. The rolling should be continued until the piece is down to 18 or 20 gauge B and S.

8. Having made sure that the bar is perfectly flat, one may cut or saw it into two or three sections.

9. Solder the sections together, as in step 1, making a bar composed of 10 to 15 layers of metal. In order to keep silver on the outside of the bar—for functional wear in the finished piece—it might be necessary to solder an occasional thin sheet to the outside surfaces of the bar.

10. File the edges even all around. Uneven edges cause cracking and separation of layers during the rolling process.

11. Roll the new bar, annealing when necessary, until it is thin enough to cut and recombine again. Laminations of 50 or more layers are possible—if strength for rolling holds out!

12. When the last combination is rolled out to 18 gauge, finally the piece may be flattened by carefully rolling or malleting or it may be formed into convex or concave contours with punches on the lead block. Be sure that the contours are simple and free of small rises or depressions.

13. The next step exercises the crea-

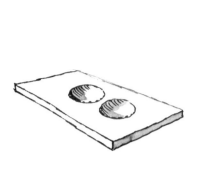

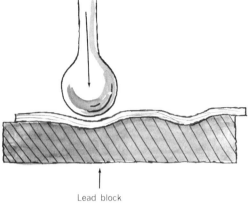

Lead block

Fig. 106

tivity of the artist in trying to anticipate the eventual result. The laminated sheet is punched up from behind with various doming or chasing tools on a lead block. The depth of the indentations must be controlled carefully since, when the protuberances on the frontal surface of the sheet are ground down, a deeply punched area might be filed completely through.

The raising and finishing steps are:

a. Forming circular depressions with a doming punch. (See Fig. 106.)

The depressions may be separate for one effect or slightly overlapping for another. (See Fig. 107.)

Make sure that the body of the sheet remains flat by an occasional blow with the mallet while the work is on the lead block. The depth of the depression should be somewhat less than the thickness of the sheet. (See Fig. 108.)

b. Liners—chisel-shaped chasing tools —may be used for punching up elongated shapes (Fig. 109A) or, if you link the blows, as in chasing a line, a straight or curved form may be made (Fig. 109B). Angling the tool creates another depres·

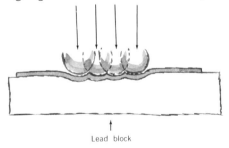

Lead block

Fig. 107

sion which results in an effective final texture (Fig. 109C).

The liners should be fairly blunt and have rounded corners so that they do not cut through the sheet. (See Fig. 110.)

c. Depressions made by doming punches may be lightly indented from the other side to form the shape shown in Fig. 111.

d. All of the raised areas on the top surface are filed down to the basic level

Fig. 108

of the sheet. This exposes the layers of differently colored metals in turn. If the top layer is to remain silver, be careful not to file too far. Slight depressions which are not wanted may be removed by light peening from behind with the domed end of a chasing hammer.

e. The file marks are removed by stoning and the surface finished by sanding with fine emery paper. A high polish is not as effective as a slightly matte finish, since the excessive reflections hide the color contrasts of the exposed rings and whorls of the various metals.

f. The unit may now be combined with other elements of the piece of jewelry or it may be the only element used.

g. The final coloring, in a potassium sulfide solution or another formula, affects each metal or alloy differently, bringing out the richness of the process even more.

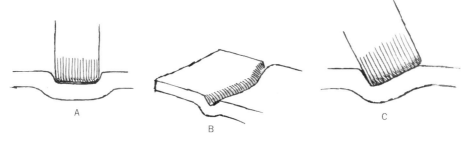

Fig. 109

h. If the completed form is dipped into a 50% nitric acid bath, the copper and copper alloys may be etched somewhat, leaving alternating raised or lowered textures.

A variation of the above technique is developed in the following manner. Five or six layers of thin metal of various colors are soldered together as in step 1 of the foregoing technique. The sheets should be 22 or 24 gauge since some-

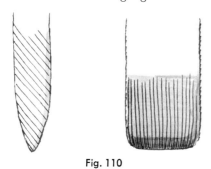

Fig. 110

what less rolling must be done. A sheet of 18 gauge metal forms the bottom or base of the piece.

With chisels, gravers, drills, or grinding wheels a variety of circular or irregular depressions are cut into the first two or three layers.

The sheet is rolled through the mill until the depressions are flattened out, leaving wide or narrow markings depending on the direction of rolling. The sheet may be etched or colored as in the previous process.

ENAMELING FOR JEWELRY

Much has been written in recent years about the art of fusing a colored, glasslike substance to metal under high heat. Contemporary uses of this ancient form of decoration are constantly being devised, and new means of fusing enamels to new alloys have been developed which give enamels far greater range than in earlier times. In this section, only the techniques applicable to the enrichment of jewelry will be described since a comprehensive discussion of the medium would make a book in itself.

Enamel consists of proportions of flint, lead oxide, soda, and potassium hydroxide. This forms the *frit,* or enamel flux—a clear, colorless glass form. With different proportions of lead and potassium hydroxide, this frit may fuse at higher or lower temperatures.

Metallic oxides are added to the frit to produce the many colors available in enamels. The addition of tin oxide makes the colored enamel opaque or translucent, depending on the proportion. In general, opaque colors fuse at a somewhat higher temperature than do the transparents.

Enamels may be fused to many metals, some of which must first be plated or treated in a way to insure a good bond of enamel and metal. The techniques of

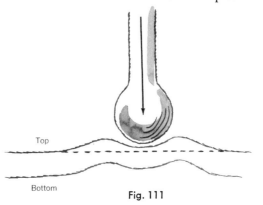

Top

Bottom

Fig. 111

105

application and fusion are similar for all of the ways in which enamel may be used.

The Enamels

The basic consistency of enamels is in sheet or lump form. This must be ground to a fine granular texture with an agate or mullite mortar and pestle.

The ground enamel must be thoroughly washed to remove impurities. At the outset of washing the water is quite milky, but when this milkiness disappears the enamel is clean enough for fusing. It may be dried on a thin sheet of metal and stored in tightly stoppered jars or it may be used immediately in its wet, pastelike consistency. A few drops of a gum solution, such as tragacanth or one of the proprietary solutions, enables the moist enamel to adhere to metal surfaces until fusion takes place.

Enamels may be purchased already ground and in a variety of grit sizes. When ground, they tend to decompose much more rapidly than do lumps, so it is best to grind the small amounts used in jewelry when needed.

Decomposition shows as white flakes and milky patches in colored transparent enamels after fusion.

The Metals

The most appropriate metals for enameled jewelry are gold (a special enameling gold is available), fine silver, and copper. Sterling silver, though much more durable than fine silver, tends to discolor transparent enamels and, when soldering is necessary, cannot use the high-melting solders necessary to enameling.

Because fine silver, even after being enclosed in the hard, glasslike enamel, is very soft and easily bent, it must often be set in sterling silver for strength. The enameled portions may be set in the manner of gem stones in bezels or other holding devices.

All of the metals must be annealed and completely free of oxides and dirt to insure complete fusion of the enamel. If the object is pickled after annealing and kept under water until ready for the application of enamels, it remains quite clean.

In some cases, a highly polished surface on the metal is necessary to the enameling technique. After the annealing and pickling process, the surface is burnished with a highly polished agate or steel burnisher without any of the polishing compounds.

A grease- and oil-free surface may be enameled even if it is heavily oxidized, with rather rich and interesting results. A clear enamel flux combines with the metal oxides to form a range of browns, reds, and greens.

Application

The enamel may be applied to the surface in a number of ways. The methods used on small surfaces are as follows:

1. Dry the wet enamel to a paste consistency by holding the edge of a blotter or cleansing tissue to it to absorb excess moisture. With a small, polished spatula —made from a 5″ length of 12 gauge copper or silver wire—a small amount of the paste is inlaid or spread over the metal surface. The spatula is also used to spread the layer evenly.

2. The enamel paste is applied to an area which has been moistened with a dilute gum and water solution. Prepare the solution by dissolving a small amount of powdered gum arabic or tragacanth in wood alcohol and add enough distilled water to make a very thin solution. A mixture which is too thick may cause discoloration of the enamel after firing.

3. Dry enamel is sifted through a fine mesh screen onto a surface coated with the gum solution. The solution should not be too dry before the sifting or the enamel will fall off. After one layer of

enamel has been fired, subsequent layers may be applied onto masked portions of the object. The portions to be kept free of enamel will have been covered with paper or masking tape stencils. When the enamel-gum solution is dry, the masks may be lifted off carefully.

If enamels must fill depressions in the metal, they should be loaded into the depression until the enamel rises above the surrounding surface. After fusing, enamels have about half the volume of dry enamels. Often two or more applications are necessary to fill a depression up to the surface. (See Fig. 112.)

A pulsing heat, caused by a rhythmic touch and withdrawal of the flame, seems to fuse the enamel easily without too much danger of overheating the metal.

Before setting prepared work in the kiln or applying the torch, evaporate all moisture from the enamel. This may be done by absorbing most of it with the corner of a blotter, then completing the process by a gentle preheating at the mouth of the kiln.

If too much moisture remains, steam causes the enamel to lift away from the metal and to scatter particles all around. If this occurs, remove the work and start

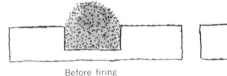

Before firing   After firing

Fig. 112

When metal of 18 gauge or thinner is used, it is best to "counter-enamel." This consists of the application of a layer of enamel to the back of the object for every layer on the front. This prevents warpage of the cooling metal after firing, which causes the enamel to crack. Doming the object slightly also helps.

Firing

Enamels may be fired (fused) in an electric kiln or with a torch. The use of a kiln assures a more evenly distributed heat and, with a pyrometer, a firing schedule may be developed which takes much of the guesswork out of this process.

A torch may be used for small pieces especially and has the advantage of allowing complete observation of the firing process. Direct contact of torch flame and enamel should be avoided if colors are to remain clear. If the work is placed on a tripod covered with a square of nickel-chromium mesh screen it may easily be heated from below.

over again after completely washing off the enamel.

Fire the work long enough so that the enamel becomes viscous. Often it will draw away from edges slightly but with continued heat will flatten out again. Overheating will again cause the enamel to crawl away from edges, as will an unclean surface.

At times, small gas bubbles work their way to the surface, and, if firing is stopped, they cause small pits. These often work out with subsequent firings and should not be filled with grains of fresh enamel, for such new applications seldom have the same color or clarity as the original.

After firing, allow the object to cool slowly. Never quench it in water or pickle for this would crack the enamel.

Finishing

The most common finish is that of the freshly fired enamel itself. This has a reflective and lustrous smoothness which may tend to confuse the eye. Sometimes a stoned finish brings out color relationships better and a slight polishing with a rouge buff restores enough of the luster without the garish reflections.

Opaque enamels have been left matte after stoning by many Oriental enamelers to further define color separations and changes.

A short dip in hydrofluoric acid etches the enamel enough to give it a translucent, frosted appearance. Hydrofluoric acid is extremely dangerous to use and the proper safety precautions must be observed! The rules are:

1. Work with rubber gloves in a well-ventilated room.
2. Do not inhale fumes of the acid.
3. Have a paste of bicarbonate of soda and water ready in case acid touches the skin. Rub on the paste gently and rinse off in cold water.
4. Keep the acid in its wax or lead container. Never pour it into a glass jar or tray.
5. Apply the acid with a cotton swab on a stick. Continuous rubbing for about 1 minute etches the enamel well.
6. Pick up the object with wooden tongs and rinse it in cold water until free of acid (test with soda paste) or rinse the entire surface on which the object was placed under a gentle stream of water. The work may be placed on a thin board for this purpose.

If some enamel areas are to be left glossy, they may be painted with melted beeswax to protect them during etching. Remove the beeswax with hot water and a rag or with benzene.

Cleaning and Soldering of Enameled Work

The metal areas may be pickled cold in a 10% sulfuric acid bath to remove the oxides of firing and may then be finished as usual.

Findings must have been soldered on with *hard* or high-melting silver solder before the first enamel firing. Since enamels are fired in preheated kilns at about 1500° F, many solders would begin to break down. The findings may be protected by a coating of yellow ocher paste if neccessary.

Delicate findings such as earring screws or clips must be applied with soft solder (lead base) after all firing has been completed, for prolonged high temperatures would soften these too much. Where counter-enamel has been used, keep a small patch of metal bare where findings must be applied in the above manner.

Decorative Processes

Most of the enameling forms have been in use for more than 2,000 years. The first enamels were inlaid in pockets or cells in metal and used much as precious stones in the decoration of jewelry. The earliest examples of enamel work are from Cyprus, Egypt, and other Middle Eastern areas. At a later date the knowledge spread to the north and to the east, where refinements and variations eventually developed.

In the Orient the Chinese, and later the Japanese, developed the techniques of *cloisonné* and *plique-à-jour* to the highest degree, whereas the Western enamelers of France and the Byzantine Empire developed the *limoges, grisaille, basse-taille* and *champlevé* processes to high perfection.

During the Renaissance, the enamelers of the French city of Limoges developed a reputation for excellence which is unmatched today. At that time enamels were often used almost as a painting medium in the *grisaille* or *limoges* technique or as integral elements in the highly complex and sculptural jewelry of the time. A fine example of sculptural enameling may be studied in the Francis I saltcellar of Benvenuto Cellini.

Most enameling on jewelry was done on gold or gold-gilded bronze, silver, or copper during this time. The transparent colors allowed the often richly carved and textured metal to shine through in the *basse-taille* technique.

The contemporary adaptations of some of the major processes are as follows:

LIMOGES OR PAINTED ENAMEL

1. The cleaned metal is first covered with a fused surface of enamel flux or an opaque color.

2. Other colors are added, either moist and pushed into place with a small brush or sifted on over stencils or masks.

3. Each layer is fired separately but may consist of as many colors as desired applied side by side. Since the granules of enamel do not fuse with each other but remain as separate spots of color, very finely ground colors must be used to form subtle transitions.

4. As many as 20 or more layers of thin color may be applied before completion, so it is important that a comparable thickness of enamel be applied to the back of the piece as the work proceeds, to avoid warping in the metal.

Since this is a painting process, the success of the piece is in proportion to the craftsman's pictorial design sense. Much mediocre work has been done recently in *limoges* enamel, since the rich color alone may be exciting, but it takes more than a permissive smattering and scattering to come up with a work that is both imaginative and of good craftsmanship. It is too easy to allow this benevolent material to dictate the final result!

GRISAILLE OR CHIAROSCURO ENAMEL

The forms are developed in white—with occasional touches of other colors—onto a black or dark blue background. The chiaroscuro effect of strong lights and darks is the basic *grisaille* quality.

1. The metal is covered with a fused layer of opaque black or a very dark blue transparent enamel. This is best sifted onto a gum-moistened surface for even distribution.

2. Opaque white enamel is ground very fine—almost to an impalpable dust if possible. This should be mixed, after thorough washing, with a thin oil or gum solution. Oil of lavender is usually used for this purpose and may be purchased from enameling suppliers.

3. The white is applied in a fairly thin layer where the form should develop. As they are fired, the white particles sink into the dark background and become gray.

4. Subsequent layers of white over all or parts of the original area become more and more opaque and white as thickness builds up. In this way very subtle modeling may be developed and may also be enhanced by the slightly raised white highlights.

Both the *limoges* and the *grisaille* enamels are best set in a framework of unenameled metal both as a protection and as a textural and color contrast.

Since enamels do not readily flow when molten, they may be applied to contoured or sculptural surfaces with no danger of distortion. Both of the above techniques lend themselves well to this treatment.

CLOISONNÉ OR CELLED ENAMEL

This ancient process may be carried out in a number of ways. The enamel is basically placed in *cloisons* or walled enclosures formed by thin flat wire.

1. The wires, usually of fine gold or silver for flexibility, are either spot soldered to the basic metal or bent to shape and placed on a surface of enamel. This may be done in two ways: the enamel, usually a thin layer of clear flux, may be fired and cooled. The wire is then put into position and held in place with a thin gum solution. The object is re-

fired, causing the *cloisonné* wire to embed itself in the soft enamel. It is important that the first layer of enamel be thin so that the *cloisons* are not immediately filled. Overfiring may cause the fine silver wires to collapse.

The second method consists of placing the wires on the clean base metal, also with a gum solution, and sifting the enamel evenly over the entire surface. The object is fired to adhere the wire.

The outer edges of a *cloisonné* panel should be re-enforced with a heavier wire (Fig. 113A), or the enamel area

2. The colors may now be placed into the cells in paste form and fired.

The enamel surface may be left slightly concave (Fig. 115A) or flush with the top of the wire (Fig. 115B), or even slightly convex (Fig. 115C). When this is done, the last layer of enamel is fused after all stoning is completed. (See Fig. 115.)

To stone the enamel or the tops of the *cloisons*—as shown in Figs. 115A and 115B—the work is placed on a chamois-covered board and rubbed firmly with a scotch stone until wire and enamel surfaces are both clean and level. This

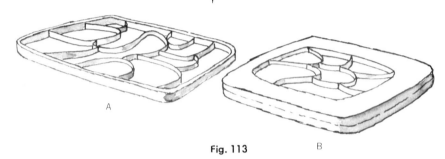

Fig. 113

may be cut out of a sheet and soldered to a backing, as shown in Fig. 113B. *Hard* solder must be used in all soldering.

The wires must touch each other to form solid cells. If gaps are left, one color might often flow into another. This may be done intentionally at times when the wire is used only as a linear device. In that case the color may even be the same in every cell.

A variation of the wire technique consists of cutting out the *cloisons* of thin sheet, perhaps 18 to 20 gauge, and soldering them to the surface of the base. (See Fig. 114.)

If the design is to be left in the interior of a form, the *cloisonné* wire should be at least twice as thick as usual. During the stoning process, one of the finishing procedures, enough vibration is developed to crack or bend fine narrow wires if they are not otherwise supported.

should be done under running water so that metal and abrasive particles float off. In the case of Fig. 115B above, the enamel may be left matte from stoning or it may be refired to a gloss. In this case, it is best to etch the enamel lightly with hydrofluoric acid to remove all stoning debris, for this would cause clouding and spotting if allowed to remain. After the etching, the enamel is refired in a hot kiln—about 1600° F—for a short time (just long enough to cause surface fusion).

PLIQUE-À-JOUR OR WINDOWED ENAMEL

This is a variation of the *cloisonné* process where there is no background

Fig. 114

110

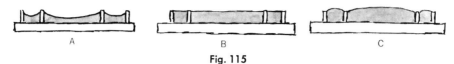

A   B   C

Fig. 115

behind wire and enamel. The effect is similar to that of a stained glass window, and, as with stained glass, must have a source of light shining through for effectiveness.

Plique-à-jour is most striking when applied to pendent earrings which allow light to come from behind.

There are basically three methods for handling this rather difficult technique.

1. Perhaps the oldest technique consists of applying the cloisonné wire to a sheet of thick mica. The wire is spot soldered only where absolutely needed since the fused enamel will eventually hold it all together. A heavier wire or enclosure of sheet metal should surround the cells since the unit will be quite delicate.

With the mica as a background, the cells are charged with transparent enamels. For the greatest clarity of color the enamel should be ground only to about 80 mesh. The larger the particles are—within reason—the brighter and clearer the color will be.

The enamel is fused and re-applied until the cells are filled to the tops of the *cloisons.*

Using the chamois-covered board, scotch stone, and water, grind both front and back so that wire and enamel are even.

Since transparency is important, it is best to etch the enamel clean and then to refire it quickly.

2. The second technique uses a sheet of copper foil of 30 to 32 gauge as a backing to which the wires are glued with a somewhat thicker gum solution than usual. The copper foil should be chemically clean to prevent discoloration of the transparent enamel.

After the enamel has been fired enough to fill the cells, the frontal sur-face is painted with hot beeswax. Make sure that all of the edges and sides of the wire are completely covered.

The entire piece is suspended in a solution of nitric acid consisting of 1 part acid to 2 parts water. This solution will quickly eat away the copper foil backing. Be careful that the etch does not attack the wires!

The work may be stoned, cleaned, and refired on a mica sheet as in the previous process.

The best method for supporting the work for the final firing is to set it on edge on a nichrome or stainless steel trivet. The danger lies in overheating to the point where gravity will cause the enamel to sag or even open up in the cells. A quick, high heat with constant observation is required! (See Fig. 116.)

It is best to avoid enamel areas larger than $\frac{1}{4}''$ squares which are not intersected with a cloisonné wire.

To key the enamel even more thoroughly, early craftsmen ran the flat wire through the round hole draw plate enough to give it a slightly curved section. The concave side was exposed as much as possible to the larger enamel areas.

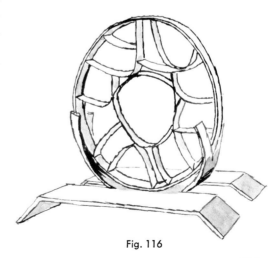

Fig. 116

3. The third and most simple technique consists of piercing and drilling through sheet metal to form the cells. These may be as close together as desired or may be separated by considerable metal surfaces.

The perforations, if no wider than $\frac{1}{8}$", may be as long as needed, but, since the most simple enamel charging technique utilizes capillary adhesion, wider areas would collapse during firing.

The inner edges of the perforations may be left rough after filing to act as a key for the enamel.

The enamel is mixed with a thicker than usual gum-water solution and floated into the cells with a fine brush.

When completely dry, the work may be supported as shown in Fig. 116 or it may rest horizontally on a standard enameling or ceramic tripod. When fired in the latter manner, it is a good idea to reverse the sides for each new firing in order to compensate for the slight sagging of the viscous enamel.

If holes appear through the enamel due to insufficient charging or excessive heat, they may often be refilled and refired.

Since some enamel usually runs over onto the flat surrounding metal it is necessary to stone and refire the piece to complete it.

Though the contrast of a large metal surface enclosing the enameled windows is very effective, the entire surface may be enameled with clear colors or flux, allowing the perforations to stand out as darker areas.

CHAMPLEVÉ OR INLAID ENAMEL

The Byzantine enamelers brought this technique to its most effective peak with a subtle use of rich enamel colors surrounded by gold or gilded metal surfaces.

The effect of *champlevé* is similar to that of inlaid metal in that both the metal and the enamel may have a common upper surface.

The oldest practitioners used very thin metal which was punched down or stamped to form the cells to be inlaid. Often these thin, quite weak units were set in bezels or prongs or were tacked to larger surfaces with gilded nails. *Cloisonné* wire was often used within the *champlevé* cells to carry out the design, forming a combination of the two techniques.

Another method of forming the depressions consisted of carving them into a thick sheet with chisels and gravers. This often left an irregular texture showing through clear enamel, though opaque enamels were most often used.

Of the many possible *champlevé* processes, two are most often used today The first is really an etching process used on copper or silver. The cells are left unpainted with etching ground and the rest, back and edges, thoroughly covered. The piece is then etched as described in the etching section of this chapter, but somewhat deeper than usual. A depth of between $\frac{1}{32}$" and $\frac{1}{16}$" is usually sufficient.

If the edges of the cell are irregular due to prolonged underbiting, they may be redefined by engraving.

Fig. 117

After the ground is removed, the piece should be annealed and pickled to clean it. Enamel is placed wet into the cells and fired. It often helps to place a drop of water into the cell before adding the enamel. This causes the granules to float into all corners and pockets. Excess water may be removed as described earlier.

The enamel may be left slightly concave or filled to the metal surface after the final firing. It may then be stoned and finished as usual.

The second technique uses two sheets of metal. One has the cells sawed out of it and the other serves as the backing. The top piece may be as thin as 22 gauge but the back should be thicker for strength. (See Fig. 117.)

Since, in most cases, the enamel will cover only a small part of the total surface, counter-enameling is not usually necessary.

Base-Taille or Sculptured Enamel

In this variation of *champlevé*, the bottom and sides of the cell are chased, carved, or ground to create textures or intaglio forms. The transparent enamel, being darker in deeper sections, carries out the modeling of high and low relief.

The enamel may cover an entire textured surface forming the entire jewelry unit, or it may fill a central cell carved into thick metal. The most complex sculptural forms were created in this manner in the Renaissance with great skill in arranging the light or dark areas.

With a flexible shaft grinder, using a variety of stones and drills, one may achieve somewhat the same effect today.

Fig. 118

Small cut shapes or granules may be fused or soldered into the bottom of a cell and covered with enamel with interesting textural effect. (See Fig. 118.)

The textural possibilities in the use of chasing tools, punches, gravers, grinders, drills, etching, and fusing are unlimited.

It is possible to combine many of the techniques described in one piece, but it is also possible to overdo what should be done with sensitivity and discrimination from the start. Because of its luster and brilliance of color, enamel is a very dominating medium and must be controlled to avoid a garish or trite result.

Additional Enameling Information

Firing

Enameled metal pieces may rest on ceramic trivets or tripods, or special holding devices may be constructed of nickel-chromium screen or sheet metal. Some of the shapes into which metal sheet may be bent to assure adequate heating from all points are illustrated in Fig. 119. It is important to avoid spilling granular enamel on holding devices since it will fuse there to contaminate future work. If spilling does occur, it is best to file or grind all surfaces that may touch the enameled work before each firing.

It is very important that the unfired enamel be absolutely dry before appli-

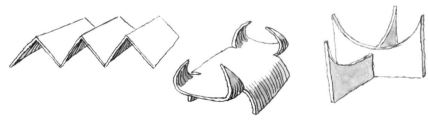

Fig. 119

cation of heat in fusing, to avoid small steam explosions which scatter the enamel. Adequate dryness may be determined in two ways. Ground enamel, because of the rough surfaces of each particle, is lighter in color when dry than when wet or fired. When the applied enamel is again the color of its dry original form before application, it is usually dry enough for firing. Many enamelers preheat the work at the mouth of the kiln or with a torch. If the enameled work is placed before a dark background, small tendrils of steam may be seen as long as moisture exists. When these disappear, firing may proceed, though it is always best to check for displaced enamel grains before finally firing the work. It is much more difficult to remove these mistakes *after* firing by scraping and stoning.

The length of the firing period is difficult to define since many types of heat are used for this purpose and also since each brand of enamel—sometimes each color—has its own fusion point.

Close observation is best. The enamel will go through several easily noticed stages during firing:

1. It becomes quite dark—though still retaining its particle shape.

2. The particles become less sharp and defined and the metal begins to turn dull red.

3. The particles fuse to each other and the total mass sinks as the enamel becomes fluid. The surface is still uneven as bumps without ridges. The enamel and the metal may now glow with the same red color.

4. If edges of the metal or walls of *cloisons* or *champlevé* depressions were not clean enough, the molten enamel may crawl away from these areas.

5. Unless the above areas were extremely dirty, the enamel, now thoroughly fired, will spread to its confining areas. The surface will be quite smooth and flat and highly reflective. At this point the work should be removed

from the kiln or torch and examined. If small gas bubbles have formed near the surface, they may be pricked with a needle and the enamel reheated for a short period to fuse these depressions together.

The work should now be cooled as evenly and slowly as possible to avoid excessive contraction of the metal. Cracks develop when the relatively inert cooling enamel cannot keep up with the more rapid contraction of the cooling metal to which it is fused.

When using copper especially, be sure that oxide flakes have not been deposited on enamel surfaces before any refiring. They would discolor both clear and opaque enamels with dark flecks. A magnifying glass or eye loupe should be used for examination.

CLEAN SURROUNDINGS

To do work of a highly professional quality, it is important to work in a dirt- and dust-free environment. Ground and lump enamels should be kept in tightly stoppered glass vials until used. The cardboard cartons and envelopes in which many enamels are shipped are not good for storage since particles of paper eventually contaminate the enamel to show up as discoloration in firing.

Mortars and pestles should be kept scrupulously clean and used only for grinding enamel. Use only mortars and pestles which are hard enough to withstand lengthy grinding without wear. Agate is the traditional material for these implements, but a more recent ceramic material, *mullite,* works very well, though it is not much less expensive.

Enamels during use may be kept in small, individual, shallow pans or in depressions in a ceramic paint tray as used for tempera painting. These should be returned—if unmixed—to their separate containers after use.

If, while a piece is being charged with enamel, work must be interrupted, it is

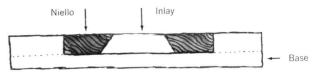

Fig. 120

best to cover the piece with a small bell jar or its equivalent to protect it from dust.

As mentioned before, enamel does disintegrate in time—especially on constant exposure to air. The opaque enamels do not suffer by this as much as the transparents since the effect is one of cloudiness and lack of clarity in color. Consequently it is always best to regrind and wash at least the transparent enamels just before use.

The decomposed enamel floats off in rinsing, leaving only clear, evenly colored particles in the mortar.

NIELLO

Niello is a mixture of various nonferrous metals and sulfur which may be inlaid by fusion into grooves or depressions in a metal base. It is an ancient technique, much used in the Middle East. Though it was often used in the decoration of armor and other metal work during the Renaissance, it has been used rarely in recent times. Today the center of this work is in Thailand, which produces large amounts of niello jewelry for export.

Because of its strong black or blue-black color, it is an effective contrast medium with light-colored metals, and may be used over large surfaces where a patina or chemical coloring would not wear well during use.

Several formulas have been handed down by such medieval chroniclers as Theophilus and Pliny, as well as the Renaissance artist, Benvenuto Cellini. These, along with more contemporary formulas, give the jeweler a wide range of possibilities and, since each craftsman must prepare his own niello, variations as to melting point are always possible.

Preparation of the Base

Though many metals may be inlaid with niello, it is most often used on sterling silver, karat gold, bronze, copper, and brass. On alloys which in themselves fuse at a low temperature, a low melting formula for niello is of great importance.

The most delicate lines or networks of crosshatching may be engraved or etched into sheet or cast metal. The depth is important since a thin layer of niello is apt to be more free of gas bubbles and porosity. On the other hand, if the engraved areas are too shallow, they may be filed away accidentally when the excess niello is removed. A depth of $\frac{1}{64}''$ to $\frac{1}{32}''$ is adequate.

Larger cells may be engraved, chiseled out, or pierced out of 20 to 22 gauge metal backed with 18 gauge or heavier metal, as described for *champlevé* enameling.

It is not necessary to angle the edges of the depressions or to *plink* the bottom surface since the niello will fuse itself thoroughly to the base metal. Smaller pieces of metal may be soldered into the interior of a cell; these should have a surface level with the surrounding metal. In early Cypriot work small metal pieces were inlaid in such a way that the fused niello held them firmly without the use of solder. In this case the metal inlay should have edges angled to a broader base. (See Fig. 120.)

After engraving and chiseling are completed, the work should be heated and pickled to free it of oxides and grease film. Keeping it in distilled water until it is charged with granulated niello will prevent further oxidation. In etched areas, all oxide residues should be cleaned out with a fine steel-bristled brush.

Preparation of the Niello

With minor variations, the ingredients of niello are mixed in the following manner:

115

1

Philip Fike, American goldsmith, has for several years concentrated on reconstructing the technique of niello. He has had notable success in mastering a process difficult to carry out with perfection and about which little is known today.

Though once a popular form of metal decoration, especially in the Middle East and Renaissance Europe, its use today, as a living art form, is virtually limited to Thailand, which has developed a considerable export industry around this technique.

Mr. Fike has experimented with many formulas for combining the elements of silver, copper, lead, and sulfur which constitute niello. In addition, he has devised new methods of application and fusion which give him greater control of the rich black amalgam.

The illustrations of both contemporary and historic niello work show it in its most effective balance; that of the soft lustrous black of niello and the reflective lighter color of gold, silver, or other metals.

It is said that niello developed as an offshoot of copper engraving as a graphic art. It is certainly true that it is a most effective means for creating a two-dimensional contrast on metal and can thus be used as a linear surface texture or form delineator.

2

3

4
5

6   7

8

1. Ring, gold and niello; Philip Fike.

2. Ring, gold, silver, and niello; Philip Fike.

3. Ring, gold and niello; Philip Fike.

4. Pair of wedding bands, silver and niello; Philip Fike.

5. Ring, gold and niello; Philip Fike.

6. Pair of shears inlaid with silver and niello. Egyptian, 3rd century B.C.

7. Hilt of Viking sword, iron, copper, and niello. Norse, 10th century.

8. Caucasian firearm, inlaid with silver, niello, ivory, and damascene. Daghestan, ca. 1825.

1. The metallic elements of copper, silver, lead, and antimony (when used) are melted together in a crucible, preferably in a muffle furnace. The silver is melted first, then the copper is added, next the lead, and finally the antimony. When all parts are completely melted, the mass should be well stirred with a strong stick of charcoal or a length of porcelain rod. Any slag forming on the surface should be removed with an iron spoon. The mixture is kept just molten until the next step.

2. In another crucible—a deep narrow shape is best—the powdered sulfur is melted. This process requires the use of a powerful ventilating fan and hood. The sulfur fumes are profuse and very acrid.

3. When the sulfur is completely melted, the molten metals are poured into the sulfur-containing crucible. The entire crucible is shaken to mix the elements and, after additional sulfur is added, the now solidified amalgam is remelted. While molten, it is stirred again.

4. The amalgam may be poured into an ingot and then broken up with a mortar and pestle, or it may be poured onto a lightly oiled steel slab and broken up with a hammer. When poured into water it breaks up into many small fragments, but tends to explode if the container is not deep and large.

The fragments are ground to about 80 mesh in a mortar and placed in stoppered containers. It is a good idea to screen the niello so that large pieces are later avoided. Some authorities claim that a powder-fine niello is more free of porosity, while others insist that larger granules fuse better.

Application and Firing

The clean metal surface is lightly fluxed with a dilute ammonium chloride solution; be sure that small depressions are not filled with the liquid. An older fluxing technique consisted of painting the metal with a barely milky borax solution made by grinding a borax cone on a slate with an excess of water.

Next the depressions are completely filled with granulated niello. On areas too narrow to be inlaid, the entire surface should be covered with niello. Since niello loses about half of its volume when melted, it should be applied thickly enough so that additional niello will not be necessary. It is always best to fuse the niello only once.

Preheat the work gently so that the water in the borax or ammonium chloride solution evaporates without disturbing the niello.

The object may be placed in a kiln at about 1000° F to fuse the niello, or the work may be placed on a thin iron or steel sheet and heated from beneath with a torch. Do not touch the niello with an open flame; this will burn it, causing roughness and pits. When overheated, niello, because of its lead content, will quickly eat into silver, causing blemishes that may not be removable. Apply the heat only long enough to cause the granules to melt and spread slightly over the surface of the object. A temperature of around 700° F fuses most niello. A small polished spatula, lightly oiled, may be used to smooth the niello into all of the depressions. Do this as quickly as possible since prolonged heat, even if not too high, may cause the lead sulfide to fuse into the basic metal.

With some of the low fusing formulas it is possible to add additional niello while the work is still hot. It may then be burnished into place with a warmed spatula.

Finishing

After the work has been air-cooled—*not* quenched!—it may be filed or scraped to expose the raised areas. The niello should remain in the depressions.

It is best to file with a worn medium-coarse file until almost through to the basic metal, and then to finish the work by stoning with a scotch stone under water. Do not use mechanical cutting or buffing since the softer niello will wear away faster than the surrounding metal, losing the effect of a continuous surface.

After the stoning, the surface may be sanded with fine emery paper wrapped around a flat stick, and polished with a felt hand-buff and rouge.

Variations

Interesting variations of the basic niello process consist of mixing gold, copper, or fine silver filings or scrapings with the granulated niello before fusion. These are filed, stoned, and polished at the same time as the niello, causing an interesting color and texture. The tendrils of metal formed in drilling through silver, gold, or copper may be aligned in parallel rows in a niello-filled depression and, after stoning, present a repeat pattern of shapes.

When niello is applied to more than one plane of an object, it may be mixed with a small amount of a gum solution to hold it in place. The work must be heated in a hot kiln so that no direct flame touches the niello. The work, held in locking tweezers, must be rotated rapidly enough to prevent having the molten niello flow out of the depressions. It is best to use a somewhat greater amount of niello for this process so that enough will always remain in the depressions even if some sagging takes place. This process is generally very difficult to perfect since molten niello has little capillary adhesiveness.

Formulas

All metals used must be free of oxides and grease and, when silver is used, it should be fine rather than sterling silver.

1. Pliny	Silver	3 parts
	Copper	1 part
	Sulfur	2 parts
2. Cellini	Silver	1 part or 1 oz.
	Copper	2 parts or 2 oz.
	Lead	3 parts or 3 oz.
	Sulfur	Half a handful
3. Augsberg No. 1	Silver	1 part
	Copper	1 part
	Lead	2 parts
	Presumably added to about ½ cup sulfur	
4. Augsberg No. 2	Lead	1 part
	Mercury	1 part
	Sulfur	1 part
5. Rucklin No. 1	Silver	3 parts
	Copper	5 parts
	Lead	7 parts
	Sulfur	6 parts
	Ammonium chloride	2 parts
	Borax	24 parts
6. Rucklin No. 2	Silver	1 part
	Copper	2 parts
	Lead	4 parts
	Sulfur	5 parts

7. Persian Niello	Silver	15.30 Gm.
	Copper	76.00 Gm.
	Lead	106.00 Gm.
	Flowers of sulfur	367.00 Gm.
	Ammonium chloride	76.00 Gm.

8. Modern French Recipe	Silver	30 parts
	Copper	72 parts
	Lead	50 parts
	Sulfur	384 parts
	Borax	36 parts

9. Recipe described by H. Wilson in "Silverwork and Jewelry"

	Fine silver	6 Pennyweights (dwts.)	.300
	Fine copper	2 dwts.	.100
	Fine lead	1 dwt.	.050
	Fine flowers of sulfur	$\frac{1}{2}$ oz.	.500

10. Theophilus	Silver	2 parts
	Copper	1 part
	Lead	$\frac{1}{2}$ part
	Sulfur	Some (an excess?)

11. Karmasch (Russian— 18th Cent.)	Silver	15 Gm.
	Copper	90 Gm.
	Lead	150 Gm.
	Sulfur	750 Gm.

12. Contemporary Russian	Silver	$1\frac{1}{2}$ oz.
	Copper	$2\frac{1}{2}$ oz.
	Lead	$3\frac{1}{2}$ oz.
	Sulfur	12 oz.

13. Bolas No. 1	Native antimony sulfide, finely ground	2 parts
	Native lead sulfide, finely ground	1 part
	Powdered sulfur	8 parts

14. Bolas No. 2	Silver	2 parts
	Copper	4 parts
	Antimony	1 part
	Lead	1 part
	Sulfur	

15. Spon

| *First Crucible:* | Flowers of sulfur | 27 oz. |
| | Ammonium chloride | $2\frac{3}{4}$ oz. |

Second Crucible: (Poured into first after fusion)	Silver	$\frac{1}{2}$ oz.
	Copper	$1\frac{1}{2}$ oz.
	Lead	$2\frac{3}{4}$ oz.
	Ammonium chloride	Trace

16. Heinrich

Silver (either sterling or fine)	1 oz.	
Copper	2 oz.	
Lead	3 oz.	
Sulfur	6 oz.	

Mr. Leonard Heinrich, Armorer for the Metropolitan Museum of Art, has repaired many valuable historical works with formula No. 16 using the following technique:

a. Melt the silver and the copper in a crucible.

b. Add the lead and stir with a charcoal stick.

c. Pour this mixture into a larger crucible containing the sulfur. Stir and cover. Let cool.

d. Melt again.

e. Pour into a bowl of water, preferably through a screen which breaks the niello into small grains.

f. Wash in cold water until the water remains clear.

g. Grind in an agate mortar until as fine as possible.

h. Place in a fine linen cloth bag and shake so that only the finest grains will come through. Take the coarser grains left in the bag and regrind. Repeat until all is fine. Finely ground niello melts more rapidly and uniformly, leaving fewer pits.

i. Wet the work with Handy Flux (a paste flux) thinned to the consistency of milk. Make a paste of the niello with a saturated ammonium chloride-water solution. Apply the paste to the parts to be covered as you would enamel.

j. Place the work on an iron plate or sheet and heat from below with a torch. About 700° F is sufficient to fuse the niello. Do not bring the metal to a glow or fire it more than twice or the niello will pit the work.

k. When the work is cool, scrape off the surplus niello and polish with water of Ayr stone and a burnisher.

If using a motor-driven buff, use a large-surfaced felt wheel.

17. Student experiment:

Silver	90%
Copper	10%
Antimony	1%
Sulfur	In excess

a. Melt silver and copper together.

b. Add molten mass to sulfur already melted in a larger crucible.

c. Add antimony and additional sulfur wrapped in a twist of tissue while above mixture is still molten.

d. Stir with a charcoal stick and let cool.

e. Remelt and apply as in other formulas. A niello with a higher melting point as well as a harder texture may have 1% to 2% nickel added to the above formula.

COMBINING METALS WITH OTHER MATERIALS

Metals, especially the precious metals, have a beauty of their own and do not need the enrichment of other materials unless a design statement may be made more meaningful by such a combination. The jewelry designer must always make the correct decision: to allow the metal to dominate the total concept or to place it in a position of backing or surrounding another material.

It is interesting that complex, highly sophisticated civilizations tended to use metal only as a vehicle behind or around gems, enamel, etc., while simpler cultures used the colors of gems and enamel only as accents in a larger metal treatment. Perhaps both directions are valid, but, since metal has been the

A

Link soldered to wire

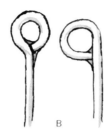
B

Wire bent into
circle and soldered

C

Wire thinned,
bent and soldered

D

U-shape soldered
into holes in
thin metal plate

Fig. 121

basis of most jewelry throughout history, it would seem logical that hiding it completely behind incrustations of diamonds and layers of enamel negates its value as a material of great beauty and personality.

The balance between materials must be decided in each case—it cannot be covered by a formula. As in all expressive actions, restraint and understatement last better and longer than do flamboyant exhibitions.

The following sections describe how nonmetallic materials may be used either by themselves or in combination with metal.

Wood

Exotic or domestic hardwoods such as ebony, rosewood, amaranth, cocobolo, walnut, maple, and many others are all usable because of their dense texture and grain beauty. Being hard, they may be sawed into thin sheets or carved in the round with little danger of splitting or chipping. Many of these woods may be worked almost like metal in that the jeweler's saw, files, emery paper, and various polishing media may be used to give them shape and finish.

Since large pieces are not necessary to the jeweler, it is often possible to purchase scrap wood from lumber companies which supply the furniture industry.

Wood may be used in jewelry in a number of ways:

Small sculptured forms with the only metal used being in the findings.

Set in a bezel as a gemlike unit.

Inlaid into the surface of a metal form.

Alternating with metal or other materials in such things as necklaces, bracelets, and earrings.

Metal inlaid into wood, the wood forming the major mass of the design.

There are a few points worth mentioning regarding the above uses of wood.

Where an independent form of wood is used, the findings must be applied so that they function well and will not be overly noticeable. For pendants a hole may be drilled at the top or just behind the top of the form. The hole should be about one and one-half times the diameter of the wire used to make the pendent loop and should be as deep as possible.

The wire loop may be made in any of several ways shown in Fig. 121.

The wire stem should be bent into a wiggle shape to fit the hole tightly and may be cemented into place with liquid pearl cement or some other hard-setting adhesive. If a large hole is impossible, the wire stem may be roughened by striking it with the edge of a rough file to achieve the same effect. (See Fig. 122.)

Very dense woods such as ebony may have the drill hole threaded and a threaded wire stem may be screwed into

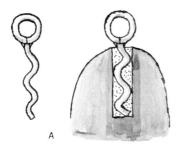

Fig. 122

place, with the use of a little cement in addition.

For heavy hanging objects, the hole may be drilled at an angle, and the wire stem may be bent to a hooked shape. (See Fig. 123.)

Pin assemblies and cuff link or earring findings may be attached by soldering them to a small 20 or 22 gauge sheet of appropriate shape which is fastened to the wood with pegs and a good contact cement. The wire for the pegs should be of the same size as the hole; 18 gauge wire fits the hole drilled by a No. 60 drill bit very well and is strong enough for most pegging needs. It is best to score the back of the metal plate and to remove all grease, wax, or oil from the wood at the point of contact. This assures a good bond of metal and wood with the adhesive. In addition, most cements require that a thin layer be allowed to dry on the opposing surfaces before additional cement is used for the final bond. Pressure from clamping clothespins for 24 hours helps in most cases.

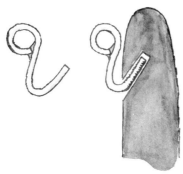

Fig. 123

Lining up the wire pegs with the drilled holes is simple with the following technique:

1. After soldering, file the ends of the pegs even and level with each other.
2. Carefully perforate a piece of tracing paper with the pegs, making the holes as small and perfect as possible.
3. Use the paper as a template to indicate where the holes should be marked on the wood for drilling.

Or you may use this technique:

1. Rubber cement the plate to the wood in its correct position.
2. Drill through the plate and into the wood as far as necessary. Use a guide on the drill bits to control the drilling depth.
3. Remove the plate. Rub off the rubber cement. Solder the pegs into the holes of the plate. After soldering, the pegs will line up with the holes in the wood perfectly.

Remember that all soldering, both hard and soft, must be completed before the wood and the metal are combined. If metal must be colored, this also must be done before combining since the coloring chemicals may discolor the wood.

When set in a bezel, the edges of the wood must be chamfered or beveled to a cabochon shape. A little contact cement, not enough to show after setting, may also be used. (See Fig. 124.)

Wood may be inlaid into metal by

Fig. 124

preparing a depression of the exact shape as the wood, but with inward-sloping edges. If the wood is a bit too large for the hole, so much the better. A few taps with a hammer on a small steel block will set the wood into the hole tightly. A spot of contact cement helps also.

It is best to cut the hole to hold the wood first and to use it as a template for the wood form. By careful sawing and filing you may achieve a very good fit.

Since some forcing of the wood may be necessary, an end grain shape is better to use than one with the grain running horizontally. (See Fig. 125.)

Both metal and wood are filed and sanded to a level surface together.

A reversal of the above process consists of inlaying the metal into the wood.

Unless one is skilled with a knife or engraving tools, cutting a bed into wood for a flat metal shape is quite difficult. The edges must be very precise and vertical so that no gaps show between the two materials. A thin piece of wood may be pierced and set in a bezel with a metal form of the same thickness used as an inlay, but it will have little strength unless fully supported at edges and back.

Thin lines of metal may be inlaid in wood by inserting thin sheets of metal into lines sawed into the wood with fine jeweler's saw blades. Complex angles and curves must be avoided because the wood may easily break when fibers are short. The metal must be rolled to the same thickness as the width of the saw cut. (See Fig. 126.)

The thin sheets are carefully bent to shape and lightly tapped into the cut. They should be wide enough to go completely through the metal. Both wood and metal are filed and sanded level.

Circles of metal may be inlaid by drilling holes into or through the wood and inserting wire of the same size as the drill bit. Both wire and sheet should be of fine, unalloyed metal, if possible, in order to allow for burnishing to fill in gaps that may occur, but with careful fitting any metal may be used.

When wood forms are used to alternate with metal shapes in necklaces, etc., they may be backed with thin metal to which the connecting links are soldered or, where thickness permits, they may be drilled to allow passage of a chain or thong. In time, such a hole will wear considerably, so it is best to insert a section of thin metal tubing as a protection.

Hardwoods may be finished by soaking them in a container of melted bees-

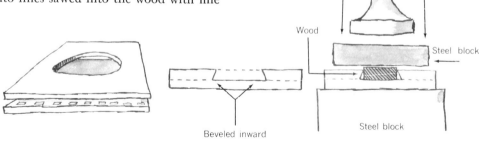

Wood

Steel block

Beveled inward

Steel block

Fig. 125

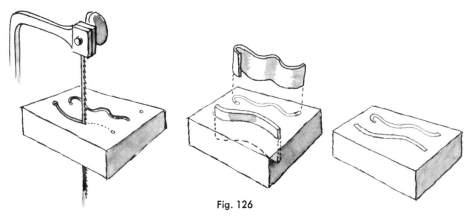

Fig. 126

wax for half an hour. They may be buffed to a soft luster by hand with a soft, lint-free cloth.

Ivory and Bone

Ivory and bone may be used in the same ways as described for wood, but they have the advantage of greater strength in working. Both hardwoods and the above materials may be polished to a high luster with emery paper and white rouge. With colored abrasives there is danger of staining the material by forcing the abrasive into the grain. Should this occur, brush the object lightly with a soft brush and a detergent solution. This raises the grain in wood and it must be resanded when completely dry.

Plastics

The many types of plastics developed in recent years have been misused so often that many artists feel that they have little inherent potential as a creative medium. This cannot be true since a material is never good or bad in itself. It can only be the intelligence behind the working of a material which may be lacking in imagination or sensitivity.

Plastics have such a variety of characteristics that some types can always be found that are suitable for serious decorative art forms. Using only two of the several basic qualities—the hardness and the rigidity of Plexiglas and nylon and the liquid-forming potential of the poly-

esters—an imaginative craftsman may find an infinity of worth-while jewelry applications.

Perhaps one of the reasons that plastics suffer a bad reputation is that they have often been used to simulate other materials. There are plastics made to simulate marble, glass, wood, metal, and so forth.

Taken as pure basic materials, the colored or clear plastics are no more obtrusive than any other basic substance, and may be used as decorative media in their own right.

As rich as the solid plastics may be as decorative materials, the liquid forming plastics have a far greater potential. They may be cast into forms and other materials may be imbedded in them while liquid and, after hardening, appear as floating images in a clear, transparent substance.

Some of the countless materials which could be imbedded in plastic for jewelry are: delicate wire and sheet forms —too fragile to stand normal wear; plant forms such as leaves, grasses, bark and other fibers; fabric; wire mesh; insects; rough and polished stones; complete constructions of precious metal to be enhanced by the enclosure within transparent plastic.

By laying in a layer at a time, a sense of dimension may be developed as in no other way. Solid forms may overlap or curve around each other. Translucent forms may show others far beneath them in the plastic.

There are steps to follow in the use of polyester plastics which are similar in all types. Each manufacturer supplies information regarding proportions of resin to catalyst, setting temperatures, setting times, and so forth. Some additional information for specific uses follows:

Mold Construction:

1. Construct a form into which the plastic will be poured. This may be made of thin metal, such as shim brass, heavy paper or cardboard having a glossy surface, or plaster of paris. Small glass bowls or trays may also be used if coated with a separating film for eventual removal of the hardened plastic.

Porous materials should be sealed with a thoroughly dried coating of shellac and also coated with a separating film.

Thin metal forms may be built on a sheet of glass and held in position with an outside fillet of plasticene. (See Fig. 127.)

Fig. 127

2. The polyester is prepared by measuring very specific amounts of resin and adding measured amounts of hardener. At this time also a number of special dyes for liquid plastic may be added. Small amounts of transparent dye may be added to create faint color tones or opaque dyes used for deep, rich color.

A sheet of colored plastic, transparent or opaque, may be cast, hardened, and sawed into small units which may then be inlaid in clear or colored plastic in a mold.

3. After the plastic unit, with additions in place, has begun to jell, it must be *cured*. With small castings (those not

more than ¼″ deep), heat is usually used during the curing process, which lasts no longer than 1½ hours.

With larger castings (those 1½″ deep or deeper), the thickness must be built up quite gradually by casting and jelling no more than ½″ layers at a time. In addition, some of the hardening chemicals should be reduced and the jelling should take place at room temperature. Curing starts only after all chemical heat has reached room temperature—this might take as long as 24 hours. The curing by heat is prolonged by slow rises in temperature over a 6-hour period. Where large amounts of metal are imbedded, the casting must be cured at room temperature for a week or more.

Cracks and fissures sometimes develop, radiating out from inlaid metal forms. These are caused by speeding the jelling and curing processes too much. To retard this on large castings, it might even be necessary to immerse the casting in ice water for a time.

4. After curing is complete, it is often necessary to shape, refine, and polish the object.

Shaping may be done by sawing, filing, and sanding, using metal-working materials in each case.

Cutting or buffing may be done with tripoli compound on a fairly slow wheel. Do not apply too much pressure or buff too long between cooling dips in water since the plastic surface may soften and absorb the cutting compound.

Final polishing may be done in the same manner by using white rouge as the polishing medium. The plastic surface may finally be protected somewhat by waxing and polishing and—should it become badly scratched during use—may be repolished and waxed to maintain its transparent clarity.

USING THE ACRYLICS

Most of the acrylics are shaped and finished in the same way. They may be sawed by machine or hand saws. Since considerable heat is developed in sawing, it is possible to weld plastic chips to the saw blade during sawing. To prevent this, use long strokes to clear the chips, use no excessive pressure, and make occasional pauses to let the plastic cool. Filing and buffing with tripoli compound refines these plastic forms very well. White rouge on a soft, un-stitched-flannel polishing wheel brings up a very fine lustrous polish.

Twist drills used for soft metals work well in drilling the acrylics and cut cleaner, more transparent holes if they are lubricated with beeswax. Drilling pressure should be light, steady, and slow —about $2\frac{1}{2}''$ per minute.

Acrylics, such as the several forms of Plexiglas, may be formed by heating. Hot water does not produce high enough temperatures for forming, so an electric oven, capable of heats from 250° F to 300° F, must be used. Infra-red lamps may be used, but they will easily overheat the surface of a sheet before the interior is soft enough for forming.

Two methods of achieving a desired form are practical for the craftsman.

1. The Plexiglas is heated to about 275° F, quickly removed and bent to shape by hand and immediately covered with a soft insulating blanket to slow down the cooling time. If cooled too quickly, the piece may assume its original shape. Use insulated gloves or holding clamps to bend the sheet to prevent burning the fingers!

2. A mold may be constructed of wood—suitably protected with a coat of synthetic rosin varnish—or made of cristobalite casting plaster.

The sheet is placed over this mold and the two placed in the oven together. As the plastic softens it sags into or over the mold, following simple contours quite accurately.

Again, the plastic must be cooled slowly by turning off the oven with the piece left inside or by insulating it properly if removed.

There are several cements to be used on one or more of the plastics. The polyesters are best cemented with small amounts of the same material, while the acrylics usually require a solvent which causes a welding action. For Plexiglas either ethylene dichloride or methylene chloride acts as a cementing solvent.

There are a number of books and manufacturers' pamphlets which describe in detail the many variations and complexities of the plastics. These are listed in the Bibliography.

GEM STONES

There are hundreds of minerals which possess the beauty and durability necessary to a gem stone. At various times in history, a mineral, because of its color or rarity, became precious to the jeweler of the day. At a later time fashion may have relegated this mineral to obscurity, putting another into its place.

Stones considered precious today usually possess a combination of richness in color and real or artificial rarity. The diamond is by no means a rare stone—its high cost is carefully maintained by mining and diamond-cutting syndicates.

Most of the *precious* stones are clear or translucent and are cut into facets to exploit their refractive and reflective qualities. Diamond, emerald, sapphire, ruby, and some opals are considered, by most gemologists, to be *precious,* while all other stones are considered *semiprecious.*

The semiprecious stones are too numerous to list completely, since their value varies with the user's attitude. If a stone has a color or texture that is pleasing it may be used in jewelry even though it is very common.

The following list contains many of the better known gems, both precious and semiprecious. The listing is divided by colors which hold true in most cases, though a mineral may often have several color phases. These will be mentioned after the stone's listing in each case.

The symbol "O" designates that the stone is opaque. "T" means transparent and "TL" means translucent.

Some gems are very hard, are tough, and resist wear well, while others may be soft and easily shattered. One scale of hardness for minerals which is often used is called the Mohs Scale of Hardness. This scale ranges from 1, the softest mineral, to 10, the hardest mineral. The list of gem stones also contains the numbers of this scale.

Color	Mineral	Hardness	Quality
Black	Agate (all ranges from white to brown to black)	7	O; TL
	Hematite	6½	O
	Obsidian	5	O; TL
	Onyx	7	O; TL
Purple	Amethyst	7	T; TL
Blue	Aquamarine	7½	T
	Azurite	4–6	O
	Lapis lazuli	6	O
	Sapphire	9	T; TL
	Sodalite	6	O
Green	Agate (moss agate)	7	O; TL
	Alexandrite (in daylight)	8½	T
	Amazonite	6	O
	Aventurine	7	O
	Bloodstone	7	O
	Chrysocolla	4–5	O; TL
	Chrysoprase	7	TL
	Emerald	8	T
	Jade (also white, gray, brown, black)	6½–7	O; TL
	Malachite	4–5	O
	Olivine	6½	T
	Periodot	6½–7	T
	Tourmaline (also pink)	7–7½	T
Yellow	Agate	7	O; TL
	Topaz (corundum)	8–9	T
	Topaz quartz	7	T
Red	Agate	7	O; TL
	Alexandrite (under incandescent light)	8½	T
	Carnelian	7	TL
	Coral	3½	O
	Garnet	6½–7½	T
	Jasper	7	O
	Ruby	9	T; TL
	Spinel (also pink, blue, green)	8	T
Pink	Rhodochrosite	6½	O
	Rhodonite	6½	O
	Rose quartz	7	O; TL
	Tourmaline	7–7½	T
Brown	Agate	7	O; TL
	Amber	2–2½	O; TL
	Cairngorm	7	TL
	Smoky quartz	7	T
	Tiger eye (often dyed red, blue, and green)	7	O
Opaque white	Pearl (with green, red, pink, brown, blue, and gray tints)	3–4	O
	Quartz	7	O

Color	Mineral	Hardness	Quality
Clear or translucent white	Agate	7	TL
	Diamond	10	T
	Moonstone	6	TL
	Quartz (rock crystal)	7	T
	Spinel	8	T
	Zircon	7½	T
Opalescents	Labradorite	6–7	
	Opal	6	

Stones of a hardness of 6 or less should be set in such a manner that they will be protected from abrasion and shock.

There are two basic shapes into which stones are cut. The oldest form—little changed for several thousand years—is the *cabochon* cut. This refers to any stone which has not been cut into facets to reflect or refract light. Usually these stones have an evenly rounded top, but this may vary considerably.

The second type of basic cut is that of the *faceted* stone. These stones have mathematically precise planes ground into the surface which reflect light as well as break it into spectrum colors. Diamonds, if cut as cabochons, are dull, colorless, and glasslike. However, cut with facets, they assume sparkle and brightness caused by the above actions of light.

Recently raw gem materials have been used in jewelry with dynamic effect.

Often these fragments—or *roughs*—are refined by tumbling them for many hours in abrasive-filled rotating containers. The effect is that of a highly polished water-worn pebble which still maintains its original fragment shape even though smooth and flawless on the surface.

Some of the more common cabochon shapes are shown in Fig. 128.

Some of the basic facet-cut stone shapes are shown in Fig. 129.

Stone Setting

THE BASIC BEZEL

In general, the faceted stones are cut with many flat planes to reflect light. They are set with prongs or in crowns to allow light to penetrate the stone easily. Cabochon stones, rounded or flat in section, are most often set with bezels or inlaid into metal.

Of the many stone-setting processes,

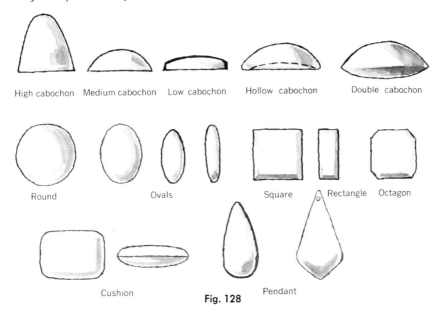

High cabochon   Medium cabochon   Low cabochon   Hollow cabochon   Double cabochon

Round   Ovals   Square   Rectangle   Octagon

Cushion   **Fig. 128**   Pendant

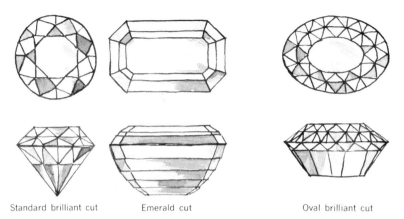

Standard brilliant cut   Emerald cut   Oval brilliant cut

Fig. 129

the bezel is effective and the least complicated. The bezel material is usually a thin strip of fine silver or gold shaped to fit the stone closely. This collar is soldered to the object and, after the soldering, coloring, and polishing have been completed, it is pushed against the side of the stone to hold the stone firmly. (See Fig. 130.)

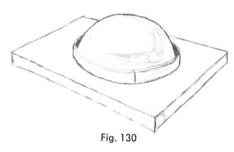

Fig. 130

Following are the steps in the construction of a simple bezel:

1. Determine the width of the bezel strip. The strip must be wide enough to go over the inward curve of the stone to just the right height. In Fig. 131A the bezel is too high. When burnished around and over the stone it would hide much of the stone, and the excess metal would wrinkle and fold. In Fig. 131B the bezel is too low. Not enough metal could be burnished over the curve of the stone to secure it well. In Fig. 131C the bezel is of the correct height. It is low enough so that a minimum of metal will be seen from the sides and the top. The bezel metal itself is 26 or 28 gauge fine silver or gold.

2. After a strip of bezel silver of the correct width has been cut, it is carefully bent around the stone at the base. A mark is made where the ends overlap. The strip should fit around the stone tightly and be without kinks. (See Fig. 132.)

3. Cut the strip a bit smaller than the mark would indicate. This takes up the slight slack caused by overlapping the ends. It is better to have the strip too short than too long.

4. The two ends of the strip must be filed to fit perfectly. If the cutting shears have twisted the metal, a firm pressure with flat-nose pliers will align the ends.

5. Fit the ends tightly together by press fitting. This is done by moving one end over the other several times so that tension will hold the edges together.

6. Use a small flame to melt one small

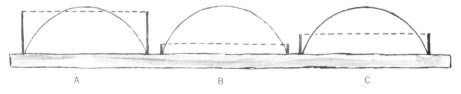

A   B   C

Fig. 131

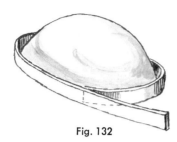

Fig. 132

piece of *hard* solder into the join. Place the solder on the outside of the bezel. Since the bezel is of malleable fine metal, it need not be formed perfectly before soldering. It may be shaped to a flat oval for stability during soldering and later brought to the shape of the stone. (See Fig. 133.)

Apply the small flame from the top. If the solder seems to flow to one side of the join only, quickly focus the flame on the opposite side. This often draws the solder into the join.

7. Pickle the bezel and remove all flux in hot water.

8. Fit the bezel around the base of the stone. It should be just tight enough to allow the stone to drop through without tilting. If the bezel is too tight, it must be stretched. This may be done by inserting a smooth rod of steel of the right diameter and rolling the bezel back and forth on a hardwood block. Apply pressure to stretch the metal. Check the fit often to avoid stretching the bezel too far. (See Fig. 134.)

If the bezel is too large it must be cut open, have a section removed, and again soldered. Make two cuts, one on each side of the old join.

Hard solder

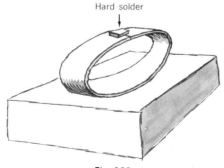

Fig. 133

9. The bottom and the top edges of the bezel should be parallel. They may be filed true with a fine flat file—number "0" cut—or rubbed flat on emery paper. Remove all burrs of metal on the edges. Burrs remaining on the top edge cause a frayed surface after the stone is set and burrs on the bottom edge might cause difficulty in soldering.

10. A hole, the shape of the stone and about three-fourths of its diameter, is cut into the backing metal, if the stone is transparent or translucent. It is best to drill a small hole through the backing even for opaque stones. This simplifies removing the stone if it should become necessary.

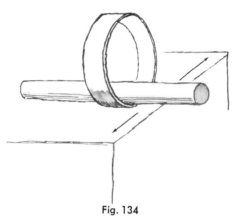

Fig. 134

11. The finished bezel is placed around the hole and soldered to the backing. Easy solder is used in most cases and is placed on the inside of the bezel in a symmetrical pattern. It is not necessary to wire the bezel into position since gravity will hold it in place. Use as small a flame as is practical. Concentrate it around the outside of the bezel in such a way that all the solder will flow at the same instant. Correct heat should draw it to the inner bottom edge. Do not lean the solder pieces against the bezel. Check constantly that the solder has not moved during heating. (See Fig. 135.)

If the stone has a convex bottom half, the bezel must have an inner shelf or *bearing* to keep the stone off the backing

131

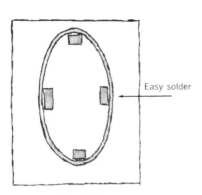

Fig. 135

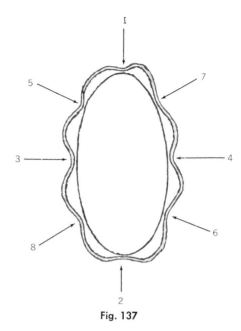

Fig. 137

metal. A stone of this shape is impossible to set firmly otherwise.

If the curve is slight, a small inner ring of wire may give the desired height. With deeper curves it is necessary to solder in an inner bezel. Some craftsmen prefer to use a manufactured bearing bezel obtainable in a variety of shapes.

An inner bearing is necessary even for flat-bottomed stones if the backing metal is curved, as in a ring or bracelet. In this case, the width of the bezel should be enough to allow the filing of a compensating curve. The inner bearing should be soldered in place before the bezel is filed to fit the curve of the backing surface. (See Fig. 136.)

Never solder the bezel into a position where it becomes impossible to burnish the edge around the stone. Do not allow the side of the bezel to be soldered to another element in the design. This

would make it impossible to burnish the top edge evenly.

A bezel may be asymmetrical in shape to fit an unusual surface or stone shape but only enough bezel metal should be used to hold the stone securely. Excess bezel material wrinkles easily in burnishing.

12. The stone is usually set after the whole work has been colored and polished. The burnishing may be done with a small curved burnisher or with a stone pusher. In both cases, the bezel is pushed against the stone at opposing points in at least eight places before a final even pressure all around the stone is applied. Fig. 137 is a diagram of the beginning pressure points. The pressure points are

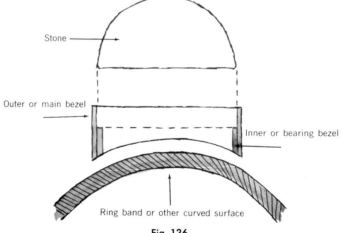

Fig. 136

132

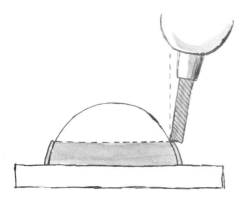

Fig. 138

opposed in sequence to keep the stone centered. After the bezel is crimped inward at eight points, the stone pusher is moved slowly around the circumference with short, even, overlapping pressures.

After the bezel lies smoothly around the stone, the burnisher may be used to remove any tool marks. Have complete control of the burnisher at all times since a slip here may cause considerable damage.

The top edge may be made smooth, and the stone tightly set, by running the pusher around the edge once or twice. (See Fig. 138.)

THE REVERSE BEZEL

Cabochon stones may also be set *into* the backing metal in a number of ways.

In one technique the hole through which the stone penetrates is cut and filed to the right size and angle to fit the stone securely. (See Fig. 139.)

If the stone is transparent or trans-lucent, a narrow bezel is soldered on the reverse of the backing metal, as in Fig. 139B. This is burnished over sharply to press the stone firmly into the hole cut for it. Do not use more bezel width than necessary.

If the stone is opaque, three or four tabs of fine silver, 22 gauge, may be used to clamp the stone into place. Note that in Fig. 140A the tabs are soldered only to a point *near* the edge of the hole. If soldered to the edge, the tabs could not be bent back to admit the stone.

THE GYPSY SETTING

The gypsy setting, though more complicated, uses no bezels or tabs on the reverse which might be unsightly in a design.

For this setting various drills and burrs are necessary. The first step consists of drilling a hole through the point under the center of the stone with a No. 65 twist drill bit.

A drill just the diameter of the stone and with a flat end is used to cut a bearing for the stone. This cut should only be about $\frac{1}{32}''$ deep.

Fig. 139

A tapered burr drill is used to cut a tapered hole from the reverse to give a section as in Fig. 141. The small hole first drilled acts as a guide for the following drillings.

A narrow flat graver is used to cut a shallow ring around the stone to form a thin ridge.

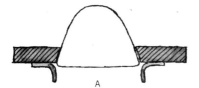

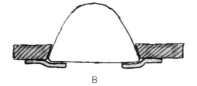

Fig. 140

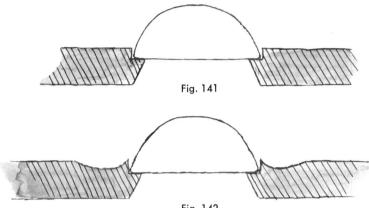

Fig. 141

Fig. 142

A roughened punch is used to hammer this ridge against and around the stone. If an invisible setting is required, the metal surrounding the engraved area may be filed and stoned level all around.

FACETED-STONE SETTING

Faceted stones may be set in any of the foregoing ways but they are shown to best advantage by prong or crown settings. When light is allowed to reach the stone from all sides as well as the top, the reflections from the insides of the facets give the stone greater brightness and color.

There are a number of techniques for making a prong or crown setting for faceted stones. Perhaps the most basic method consists of making a short tube or *collet* of 16 to 18 gauge sterling silver (a lighter gauge in gold). This tube should be of the same outside diameter as the stone and as high as the setting is to be. Tubing of the desired wall thickness may be purchased for this purpose in square or round section.

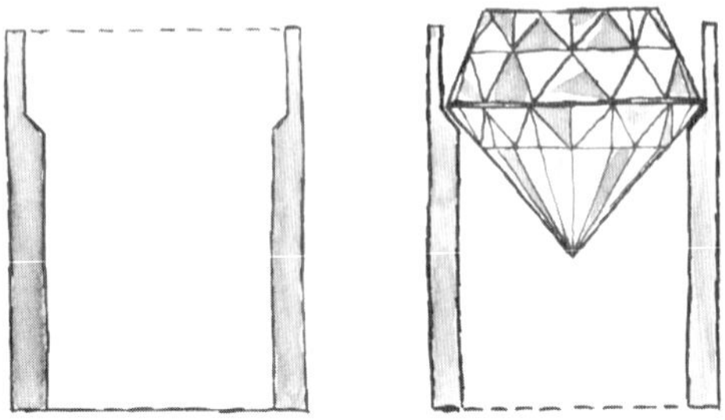

If the tube is to be constructed of sheet, it should be made in such a way that the join meets perfectly for its entire length. This may be done by overlapping the ends, while wrapping the sheet around a mandrel of the correct size, and making the join with a cut of the saw. The ends are carefully brought together and soldered with *hard* solder. True the tube up on a mandrel if necessary. True the top and bottom edges with a flat file.

Next fasten the tube to a pointed stick of the right size with melted sealing wax. The stick may be held in a ring clamp for convenience.

A bearing about $\frac{1}{16}$" deep must be cut into the inner top edge of the collet to form a seat for the stone. This may be done with a narrow flat graver, with a drum burr in a flexible shaft tool, or with a narrow scraper. Do not make the bearing too thin since this will weaken the prongs when the stone is set. At this point the collet is too small to allow the stone into the bearing. After the prongs are cut they are bent out enough to let the stone into position.

The prongs are cut by angling the saw—with a "0" blade—in such a way that one prong is formed at a time. The collet may be held on a small dowel with sealing wax. (See Fig. 144.)

The other prongs, usually of an even number, may be cut in the same way.

The result at this point looks quite

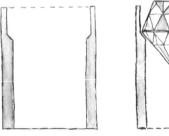

Fig. 143

Fig. 144

heavy, which would detract from the light, airy setting the stone requires. Much of the bulkiness may be avoided by beveling the edges as well as the curve at the base of each prong. If the collet is to be quite high, the base should also be notched symmetrically to increase lightness. (See Fig. 145.)

The file and saw marks may be removed with folded bits of emery paper or with a trumming string and tripoli.

After it is polished, the collet is soldered to the backing metal with small pieces of solder at appropriate points. Use enough solder to insure a strong join.

After the entire work is finished and polished the stone may be set. The prongs are slightly bent outward by pressing down between them with a round dapping punch somewhat larger than the stone.

Fig. 145

The stone is pushed into place and, while held securely with the index finger of one hand, the points of the prongs are pressed over the edge or *girdle* of the stone. Press at opposing points at all times. The stone pusher is used for this and its surface should be slightly roughened to prevent slipping.

Most craftsmen use a graver to cut the sides and the top of the prong to a point. A polished curved burnisher is used to give the final downward pressure to the points so that they cannot catch on clothing, etc. (See Fig. 146.)

Crowns and other types of facet settings may be purchased ready-made with collets to fit virtually all stone shapes

Fig. 146

and sizes, but for work that is integrated throughout all of its parts a manufactured fitting of this sort is seldom harmonious.

The Paved Setting

Another basic setting is the *paved* setting for both cabochon and faceted stones.

A hole is drilled as a guide for larger drills to follow. Next a hole is drilled with a flat drill deep enough to seat the stone securely but not so deep that much of the stone is hidden. A hole $\frac{1}{32}''$ deep is sufficient for most stones.

Another hole is drilled through from the reverse side somewhat smaller than the diameter of stone at its base. If

135

Fig. 147

the stone has a curved or pointed bottom, a tapered burr drill must be used for the last hole. (See Fig. 147.)

Next the piece is completely finished, after which the stone is temporarily fixed into the bearing with a drop of melted sealing wax. This may later be removed by a soaking in benzene.

A lining graver is used to make a series of 4 to 6 cuts $\frac{1}{16}''$ long aimed toward the edge of the stone. These cuts should angle quite steeply into the metal and stop just short of the stone. An upward tilt of the graver lifts up a small prong of metal at the end of each cut. (See Fig. 148.)

This prong is pressed down and next to the stone with a beading tool of the right size. This tool, usually purchased as a set of 12 points, consists of a round wooden handle and ferrule into which small steel rods are fixed. The rods taper at the tip and end in concave depressions of various diameters. By rocking and turning the tool gently but firmly, the prong of metal is worked into a small spherical grain which clamps the edge of the stone into place (Fig. 148B).

Many stones may be set side by side into the surface of a piece by this technique. This gives it the name of *en pavé* or paved setting.

SETTINGS FOR UNUSUAL STONE SHAPES

There are times when a stone must be set away from a backing material. In these cases structural strength of the setting is all-important since otherwise the stone might easily be lost.

Many of the finest stones seen today are cut asymmetrically or are left in a polished but irregular form, as in tumbled stones. In most cases they have no bottom plane nor do they have an even girdle around the circumference. To set these stones well tests design and construction ingenuity since the setting should, in most cases, be unobtrusive.

The now too-familiar wire wrapping of tumbled or rough stones is one simple method about which there is little to say other than that it all looks the same. Certainly this is a criticism.

There are other "cage" settings which may be related better to the total design of a work.

For rounded or plane-cut stones that are to stand free, the sketches shown in Fig. 149 may suggest additional directions.

Pearl Setting

The most secure way to set a pearl is also the most difficult. It is worth mentioning because it shows the pains to which earlier jewelers went to achieve a craftsmanlike piece of work, and there might be times when maximum security of this sort is necessary.

The pearl must be half drilled. If it cannot be purchased this way the craftsman may do this himself.

A drilling jig is simple to construct. Take a strip of 14 gauge aluminum sheet, 10" x 2", and bend it in half lengthwise. Next drill a series of holes $\frac{1}{2}''$ from the open end and through both

A

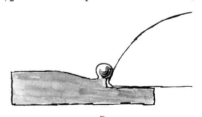

B

Fig. 148

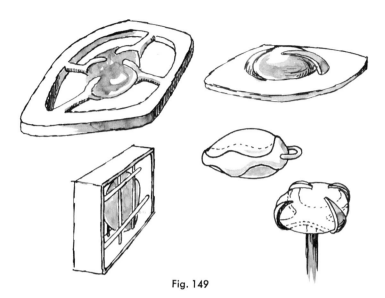

Fig. 149

halves of the clamp. Make these holes
$\frac{1}{32}''$, $\frac{1}{16}''$, $\frac{1}{8}''$, and $\frac{1}{4}''$ in diameter.
Using appropriate round dapping
punches, make a series of smooth hemi-
spheric depressions on the inside edges
of each hole. If necessary, sand and polish these depressions so that the pearl
will not be damaged during drilling.
(See Fig. 150.)

Make a sliding sleeve of a strip of
aluminum or other metal and, after
forming it around the clamp (as in Fig.
151), crimp the ends together. This
sleeve, when pushed up toward the pearl,
will press the clamp jaws tightly onto
the pearl.

Though there are special drill bits
and bow drills made for pearl drilling,
a sharp twist drill bit in an electric drill
press or flexible shaft tool will do as
well. Use a low speed and add water
occasionally for cooling. It is easy to
drill too far through cultured pearls

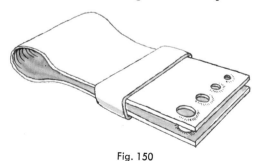

Fig. 150

since the hard nacreous coating thinly
covers a softer interior. Use a bit of
tape on the drill bit as a depth indicator
or set the bit into the chuck at the right
depth.

The peg is constructed by using two
pieces of 18 gauge round silver or gold
wire and drawing them through the
draw plate together. In time the wires

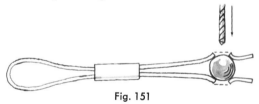

Fig. 151

become half round in section and can
be drawn to the same diameter as the
hole in the pearl. The two half round
sections of wire are soldered partially
with hard solder. It is necessary to bend
the ends apart slightly so that solder
does not flow along the entire length.
The peg is now soldered into place.
Next construct a small wedge of metal
which will be used to spread the two
wires apart when the pearl is lightly
forced onto the peg (Fig. 152C).

The hole in the pearl must be enlarged at its interior end. This is done
with a very small dental burr which is
rotated around the inside but not the
outside opening of the pearl (Fig. 152D).

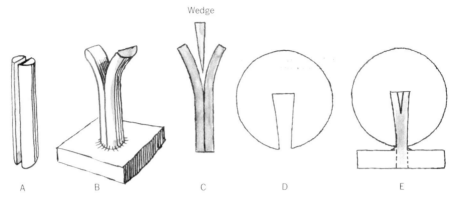

Fig. 152

A small drop of pearl cement is placed in the hole on the end of a piece of wire and the pearl pressed onto the peg with the wedge lightly in place (Fig. 152E).

If this has been done accurately, the pearl cannot fall off.

Another quite adequate technique uses a cup of thin metal into which a solid peg is soldered. The cup may be punched out on a lead block with a dapping punch the size of the pearl. The cup should not come up to more than the quarter of the height of the pearl (Fig. 153A). A small hole is drilled into the center of the cup and a wire peg—just a little thinner than the diameter of the hole in the pearl—is soldered into it (Fig. 153B). Place the solder on the convex surface of the cup. If necessary, the projection of wire may be nipped and filed off or it may be fitted and soldered into a hole in the backing metal (Fig. 153C).

After the work is polished, a drop of pearl cement is placed in the pearl hole and into the cup. The interior of the cup should be roughened with a scriber and the wire peg should be bent slightly. When the cement hardens in the hole, the bend in the peg will act as a clamp (Fig. 153D).

Pearls may also be set in cups with prongs. These may be constructed or may be purchased in a range of sizes. Pronged cups are used most often with undrilled pearls.

Pearls that are drilled through completely may be set in several ways.

1. A peg of fine silver wire is soldered to the backing metal. The diameter must be just that of the drilled hole in the pearl. Allow $\frac{1}{32}''$ of wire to project past the top of the pearl when it is in position. A beading tool of the right size is used to compress and round off the projecting end. This acts as a rivet. Do not use too much pressure since pearls are easily chipped or scratched (Fig. 154A).

2. A peg made of two wires, as in Fig. 154B, is soldered into place. The peg should fit the hole tightly. The ends

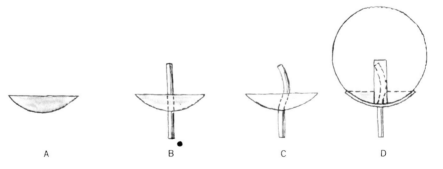

Fig. 153

138

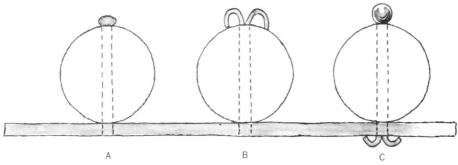

Fig. 154

should project at least $\frac{1}{16}''$ past the top of the pearl and are bent down over the pearl to hold it.

3. A peg of fine silver with a small sphere of gold or silver soldered to one end is placed through the pearl and through a hole drilled in the backing metal. The peg should fit this hole tightly. The projecting end is burnished or spread over with a burnisher or a pointed punch to rivet the peg into place (Fig. 154C).

Baroque or irregular pearl shapes may be set in beds of thin fine silver or 22K gold sheet. These must be formed to fit the contour by repoussé techniques. The addition of a peg and cement is advisable.

Sometimes baroque pearls have a flat base, in which case a bezel construction would work well.

There are several kinds of pearl cement available, but the more recent liquid types, usually a resin and a catalyst, are the easiest to use. Most cements dry slowly so it is best to clamp the pearl in position with a padded spring clothespin for at least 24 hours.

Basic Lapidary Techniques

The variety and the complexity of lapidary tools and equipment have greatly increased in recent years. Thousands of people find satisfaction in the act of creating striking gems from often uninspiring rough mineral materials.

The activity has become so important that many companies now produce cutting saws, grinding wheels, and polishing units. Other companies collect and sell rough gem materials found all over the world.

It is not within the scope of this book to go into great detail about a craft which has its own history and complexities. It should suffice to state generally the basic steps of lapidary work and to describe the most necessary equipment.

Since there are several excellent dealers in cut and polished gems—most of whom are willing to send the craftsman a good variety of stones on consignment —most jewelry makers are able to satisfy their needs directly in this way.

There will be times, however, when a stone of certain size, shape, and color must be designed for a specific piece of jewelry. Here an elementary knowledge of lapidary work becomes necessary.

Though much cutting and polishing, especially of such soft stones as turquoise and obsidian, may be done by hand with the proper abrasives, the range of possibilities is limited by both time and gem characteristics.

Lapidary units may be purchased in separate parts and may often be constructed from the most basic fittings. There is a convenience, however, to using a combined unit which includes a slab and trim saw, grinding wheels, both rough and fine, and interchangeable sanding and polishing discs or drums. (See Fig. 155.)

The steps in forming a simple cabochon stone are as follows:

Fig. 155   Courtesy, Highland Park Mfg. Co.

SAWING

The rough gem material is first sawed into usable slabs. This may be done by feeding the slab into the revolving saw by hand or by clamping it into a cradle which may move it forward automatically. The saw for this purpose is a thin disc of bronze, copper, or steel. Diamond chips or dust have been embedded in slits on the edge of the saw or fused to the edge by sintering. For most stones, a speed of 2,000 to 3,000 surface feet per minute is adequate. This means that, if a 10″ saw is used, it will require an electric motor running at 1725 rpm with a motor pulley 2½″ in diameter and an arbor pulley (to which the blade is attached) of 4″ diameter.

The saw must run through a lubricating and cooling bath. Cooling is necessary to prevent cracking the stone, and lubrication reduces wear on the saw edge. Though water may be used for this purpose, an oil-kerosene mixture is better—though messy. The solution consists of one part No. 10 motor oil to 5 parts kerosene. The sludge and debris collected after considerable sawing should be removed regularly to reduce wear on the blade.

While cutting, enough of the lubricant should be picked up by the saw so that the cut never runs dry.

Hold the work firmly so that the saw is not bent during cutting. A steady even pressure, just hard enough to continue the cut, is best. Never cut completely through a slab. Join a long cut with a short one from the opposite direction

After the slab has been cut, it may be trimmed to a usable shape in the same way.

GRINDING

The blank is next ground to its contour shape on a rough grinding wheel (100 grit). For most stones a surface speed of 4,000 to 6,000 feet per minute is best. Pulleys may have to be changed as the diameter of the wheel decreases with wear. A slow wheel is not only inefficient but also quickly forms grooves and pits which may break the gem stone.

Water should drop onto the forward surface of the wheel during grinding. If ground dry, the frictional heat may crack or permanently discolor a stone.

Unless the blank is very small, it is simpler to hold the stone by hand rather than to mount it on a dop stick. Rough grinding the stone to a symmetrical shape and to a specified outline takes considerable practice. Remember that the revolving wheel can be dangerous. It can easily cut into a finger almost before it is noticed!

140

When grinding edges it is best to avoid chipping by first grinding a slight bevel on both top and bottom edges. This bevel must, of course, be renewed as the edge is ground away to the desired circumference.

Flat surfaces may be ground on the side of the wheel. It is often difficult to cool this surface with water so the stone temperature should be checked often.

After the basic form has been ground on the rough wheel, it may be refined to remove flats and irregularities on the smooth grinding wheel (220 grit) running at the same speed.

DOPPING

Before going to the next step, sanding, the stones must be mounted on dop sticks. These may be made of 5″ lengths of hardwood doweling. A range of diameters from ⅛″ to ¾″ will handle many stone sizes. The stone is attached to the dop stick with a dopping cement made of sealing wax, stick shellac, and beeswax. This combination may be purchased from lapidary supply houses, ready to use. Sealing wax alone is an adequate cement but it tends to be brittle.

The cement is heated just to melting in a small container and the sticks are dipped to a ¼″ to ½″ depth. After allowing the cement to bond to the wood for a moment, remove the sticks and set them, cement end down, on a smooth metal surface. On cooling they should have a small, flat platform of wax to which the stone may be fused.

The stone must be heated properly to assure a strong bond. A small oven may be constructed, using a tin can into which a small alcohol burner is placed. (See Fig. 156.)

Since the bottom plane of the stone is sanded and polished first, the stone is placed flat side down on the oven top and is heated, using a very small flame to reduce heat shock. The correct stone temperature may be determined by placing a small drop of liquid shellac on the stone. When this bubbles the stone is hot enough. On sensitive stones, such as opal and obsidian, less heat should be used. The stone should be just hot to the touch.

When the stone is warm enough the prepared dop stick is heated, placed into position, and the stone lifted to a metal sheet. As the cement cools the stick should be centered and checked for the correct angle. It should be at right angles to the flat plane of the stone. The warm cement may be modeled around the base of the stone with the fingers. Dip the fingers into cold water before touching the cement or it may burn. Never allow cement to form over the edge of a stone since it would pack into sanding surfaces and greatly reduce the cutting action. Allow the cement to cool completely before going to the next step.

A well-dopped stone should look like the sketch in Fig. 157.

SANDING

Sanding may be done in a variety of ways: on circulating belts of abrasive cloth, on discs of cloth or paper cemented to wheels, or with loose abrasives on wood or metal horizontal laps.

Fig. 156

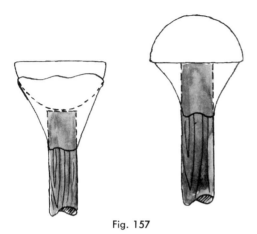

Fig. 157

The method used with the machine illustrated in Fig. 157 is the second one mentioned above. Discs of abrasive paper in three grit sizes—220, 400, and 800 —are cemented to a vertical wheel with a special rubber cement. The wheel is padded with sponge rubber under a tight cloth cover. This resilience is necessary when sanding curved surfaces. The abrasive paper should be the wet-or-dry type and should be water-cooled to avoid overheating the stone or causing the dopping cement to soften.

The sanding wheel should revolve at about 4,000 feet per minute and it should be remembered that only the outside portion of the wheel diameter achieves this speed. Speed is reduced quickly toward the center of the wheel.

In sanding, the dop stick is held with the tips of all four fingers on one side of the stick and the thumb on the other. The stick forms a line with the wrist and lower arm. By rotating the hand at the wrist, a circular or oscillating motion is given the stone so that it never rests long in one place. The pressure on the stone against the wheel varies with the hardness of the stone, but it should be uniform at all times. Sanding *with* the saw or grinding marks removes scratches faster than sanding across them.

The marks of the previous operation should be removed before going to the finer abrasive, and all abrasive debris must be carefully washed off between steps. When the 220 grit sanding is finished, the stone should be symmetrical and the subsequent abrasives used only to make it smoother.

POLISHING

After sanding, the stones and dop sticks must be carefully washed to avoid contamination of the polishing materials.

The polishing buffs may be of wood, leather, felt, or foam rubber covered with chamois.

The polishing media are:

For Soft to Medium Hard Stones	For Hard Stones
Cerium oxide	Chromium oxide
Tin oxide	Diamond powder
Levigated alumina	
Rouge	
Tripoli	
Zirconium oxide	
Linde A ruby powder	

The polishing buff is wetted thoroughly with water and a little detergent. The abrasive powder is brushed on after being mixed to a heavy cream consistency with water.

A low speed of 450 to 800 rpm is used in polishing, along with considerable pressure. Since pressure is necessary, the back of the stone should be supported with the thumb and forefinger. Use the same oscillating action of the wrist as in sanding. Check the stone surface frequently after wiping with a paper tissue. If the stone has been carefully ground and sanded, and the correct polishing powder used, the polishing step should be completed quickly.

The stone may be removed from the dop stick to expose new surfaces for sanding and polishing by heating and removing it with a quick twist. What cement remains may be dissolved off with alcohol or benzene.

After polishing, the stone should be cleaned by brushing it with warm soapy water and a soft brush.

DRILLING

Drilling holes through gem stones is quite simple, but it takes time. There are special drills made for this purpose, both electric and hand-operated. An ordinary electric drill press may, however, be adapted for drilling.

The materials for drilling are:

1. Several 2″ to 3″ lengths of thin-walled sterling silver, stainless steel, or bronze seamless tubing. Outside diameters may range from $\frac{1}{16}$″ to $\frac{3}{4}$″ or larger.

2. A small, 3″ x 3″ open box of wood or metal.

3. Plaster of Paris.

4. No. 10 motor oil.

5. Fine (400 grit) carborundum or fine diamond powder (100 to 200 grit).

The box is half filled with a thick mixture of plaster of Paris. The gem material is imbedded into the soft plaster so that the top of the stone is just level with the plaster.

After the plaster is hard and completely dry, it is sealed with a soaking coat of shellac or lacquer. This prevents the soaking up of the lubricating oil during drilling.

The abrasive powder is mixed to a creamy consistency with the oil and poured into the box to the depth of at least $\frac{1}{8}$″.

The box must be clamped securely into position on the drill press table so that the drill will constantly touch down at the same point.

The drill is prepared by lightly scoring the edge of the end with a thin-bladed knife and a light hammer. These scorings will grip the abrasive enough to cut the stone. (See Fig. 158.)

The tubing is accurately centered in the chuck of the drill press. Any wobble will cause inefficient drilling and may jam the drill itself. Make sure that the tube will come down at the right point on the stone.

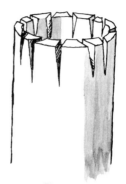

Fig. 158

Perhaps the most important action in drilling is in using a light up and down pressure of the drill. Just touch the drill to the stone and lift it again. This allows lubricant and abrasive to flood the bore after each stroke. The up and down strokes are repeated with even pressure until the drill has cut almost through the gem. At this point even lighter strokes are used until the drill goes through.

The stone is now reversed and drilled through from the other side to make the hole sides parallel.

The plaster may be carefully chipped or broken away from the stone, and the stone washed with warm water and detergent to dissolve any oil absorbed by the surface.

It is best to construct a small shield of heavy paper and masking tape at a point where the tubing enters the jaws of the drill-press chuck. This prevents abrasive particles from working up into the drill shaft where considerable damage could result.

FACET CUTTING

Many of the stones cut cabochon are also cut with facets to reflect light.

The equipment for facet cutting can be extremely simple, but it may require years of practice to achieve the skill needed for accurate and rapid faceting.

Today there are several mechanical devices available that take out the guesswork and intuition of older techniques.

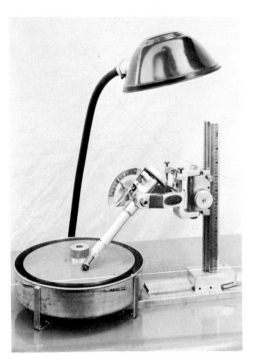

Fig. 159

The equipment consists of a horizontal lap or wheel. This may be made of hard metal such as steel, iron, or copper, and a soft metal such as lead, tin, or zinc. In addition, wood, plastics such as Lucite or Plexiglas, and wax-impregnated cloth cemented to aluminum are commonly used.

The abrasives for cutting are silicon carbide or diamond powder. Cerium oxide is often used for polishing.

The most important, and rather costly, item of equipment is the faceting head. This machine allows the lapidary to set the dop-stick-mounted stone at any angle desired. With the gauges available, he is able to plot out the precise angles of planes used in any of the many faceted cuts. (See Fig. 159.)

Faceting, even more than cabochon cutting, is a precise and complex affair which is a craft in itself.

Though the secrets of lapidary techniques were long and jealousy guarded, there are several excellent and readable texts now available on the subject.

USES OF GLASS IN JEWELRY

Stained Glass

Fragments of stained glass, available from manufacturers of stained glass and stained glass window studios, may be sawed, ground, and polished with the same techniques and materials as used in working the softer gem minerals.

The undulating surface, irregular bubble formation within the glass, and, of course, the rich color possibilities make this a very intriguing material when used as a focal point in a design. Fragments of stained glass may be set in bezels or other gem settings or may be suspended after holes have been drilled at strategic points. Stained glass is quite soft and easily shattered so holes for direct suspension must be located far enough from the edge of a form.

Small chips of stained glass may be melted into spheres and globules with a clean reducing flame or in a kiln. If fused in a kiln, the fragments should be placed on sheet mica so that no serious adhesion takes place during melting.

With care, shapes created in this manner may be attached, while still molten, to gold and silver wires leading to a number of decorative uses in a delicate design.

Since glass and metal have quite different rates of expansion and contraction, it is important to cool the combined units as slowly as possible.

Gold Glass

The art of combining metallic sheets of gold with glass has a long history. About 3,000 years ago the Egyptians began to develop a technology of fusing gold foil or leaf between layers of glass, but the skill did not reach its peak until a few centuries before the birth of Christ. From that point on, Egyptian, Greek, and Roman work developed considerably until the time of the 5th century A.D. Since that time the art has been in decline, until today only a few individuals, after painstaking research and effort, have again been able to use this very intriguing method of decoration.

As in all combinations of the two elements, metal and glass, the problems of variable expansion and contraction rates must be solved.

It is possible, through experiment, to find or develop the correct glass mixture, heating and cooling schedule, and mold material to construct gold glass for jewelry uses.

Historically, gold glass has been used in the construction of mosaic tesserae, pendent forms for jewelry and beads, but perhaps its greatest use was in the medallions inlaid into the bottoms of drinking glasses and cups.

This is another of the old decorative methods—in great part still a mystery—which has the combined attraction of rewarding research and extreme potential in jewelry-design use.

SURFACE TREATMENTS

Chasing

Chasing, as described in Chapter 3, is used in combination with repoussé to delineate and clarify a design. It may also be used in limitless ways to cover the surface of metal with an imaginative texture of indentations.

Since chasing tools come in such a variety of shapes and sizes, the creative combination of several at a time can form a unique surface. Using alternately heavy and light strokes of the hammer the same tool may be used with changing effect. (See Fig. 160.)

Simple chasing or indenting tools of limited life may easily be made of nails and spikes by filing the points flat and then filing the design imprint into the flat surface. It is best to file the surface into thin ridges rather than to leave large flat areas. Flat areas do not imprint into metal very easily and tend to bounce with the hammer blow, blurring the edges of the indentation. Fig. 161 shows an effective nail indenting tool and its mark.

Nail sets, as bought in the hardware store, are never perfectly round. This slight irregularity adds a quality to a surface that more precise punches may lack (Fig. 162A). For very small circular indentations, a set of beading tools may be used (Fig. 162B).

Matting tools used in repoussé work are designed to create an even but varied surface which is often rich in its pro-

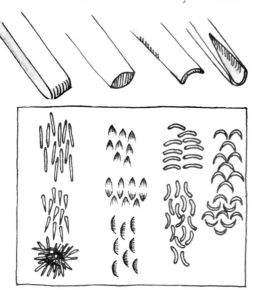

Fig. 160

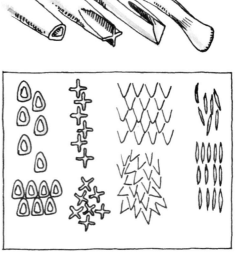

Fig. 161

145

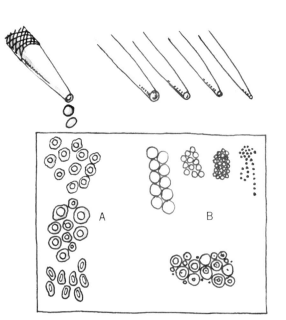

Fig. 162

fusion of raised and lowered texture. (See Fig. 163.)

Planishing

This surface treatment has been misused so often that many craftsmen avoid it altogether.

If used with discipline and moderation—with attention to the relationships between the sizes of the marks and the object—there is nothing inherently bad about this surface.

Planishing consists of covering the surface of a convex surface with a uniform and overlapping series of flat planes or facets.

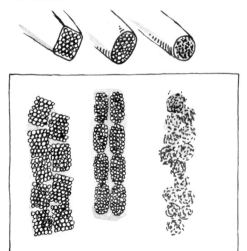

146   Fig. 163

Fig. 164

The size and weight of the planishing hammer, the flatness or curve of its polished surface, and the force of the blow all determine the shape and extent of the facet. A controlled, spaced, overlapping of small-sized planes or depressions is more effective and less garish than the effect resulting from the irregular hope-for-the-best banging which is too often seen. (See Fig. 164.)

Drilling

Very precise depressions or perforations may be achieved by using drill bits of the same or varied sizes. Cone and bud drills give shape to each depression not to be achieved in any other way. (See Figs. 165 and 166.)

On fairly thin metals a combination of drilling and indenting may be used to form craterlike projections or depressions. Metal of 20 gauge B and S or thinner is best for this. The hole is drilled first and then a tapered punch of the right shape is used to drive a small crater up or down.

Grinding

Delicate surfaces may be cut and ground into the metal surface by use of a flexible shaft tool and a variety of

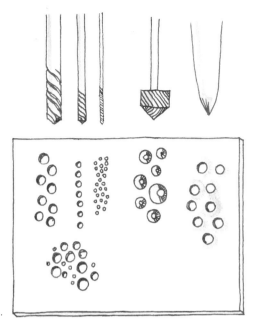

Fig. 165   Fig. 166

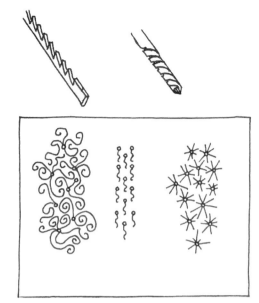

Fig. 168

small abrasive heads and dental drills. Each angle and pressure of the tool forms an effect of its own. (See Fig. 167.)

Perforation

With a fine-bladed jeweler's saw and a small drill, a linear perforation is possible which, with oxidation, may form a strong surface quality. If the perforated section is to be backed by a solid sheet of metal, only a minimum of solder should be used in combining the two sheets. Excess solder will flow into the sawed lines and greatly reduce the

effectiveness of the texture. (See Fig. 168.)

Fusing of Wire and Sheet Fragments

Small sections of wire, small fragments of sheet metal, and coarse filings of metal may be fused to the surface of a form without solder. The surface and the small fragments should be pickled clean, fluxed, and heated carefully until the surfaces of both are just molten. The result has a granular irregularity which holds color and patination readily. (See Fig. 169.)

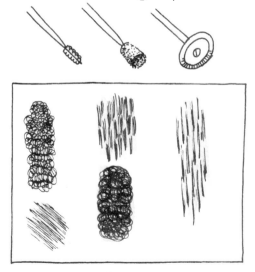

Fig. 167

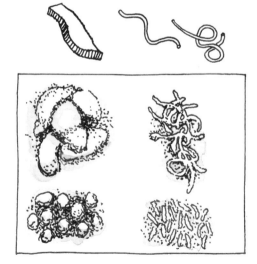

Fig. 169

This

not

This

Fig. 170

Granulation

Perhaps the most difficult jewelry technique to master today, granulation reached its highest level of excellence in the ancient world. As early as the 6th century B.C. the Etruscans were able to construct jewels of almost unbelievable delicacy and precision.

The process consists of fusing—without the use of solder—small grains (shot) to a solid surface in such a way that they are attached only at the point of contact. (See Fig. 170.)

The Etruscans and later the Greeks were able to cover considerable surface areas, as well as to align these grains to form geometric and figurative designs, using gold grains as small as $\frac{1}{160}''$ in diameter. The effect of a surface evenly covered with grains of equal size is that of a frosted shimmer of great individuality.

Since these early craftsmen had no optical aids such as magnifying glasses, they employed children for this work, and it is said that their eyesight was usually permanently damaged by the age of 10 or 12.

The art of granulation fell into disuse in early Christian times, and only at a few points in later history were attempts made to reconstruct the process.

Old records of the technique are vague and often contradictory. Often the chemicals referred to had ambiguous names virtually invented by the chronicler who, in most cases, had only hearsay knowledge of the technique. From the writings of the 12th century monk Theophilus come the most accurate accounts of granulation, and it is from these that modern goldsmiths evolved new approaches to the process.

The construction of the grains is not difficult. Though gold has most often been used, silver, platinum, and alloys of these metals have also been treated in this manner. If the grains are to be fairly large, $\frac{1}{64}''$ or more, they may be prepared by cutting fine wire into measured lengths. Smaller grains may be made by filing a sheet of metal with coarse or fine files and melting the filings into shot.

In order to melt the metal efficiently, a high, rather narrow crucible must be used. Powdered charcoal is first placed in the bottom of the crucible to the depth of at least $\frac{1}{2}''$. Over this is placed a loosely sifted layer of filings or wire snippets. Another layer of charcoal is followed by more metal particles, and so on, until the crucible is filled.

The filled crucible is placed in a muffle furnace and heated to at least 1900° F. The crucible is allowed to cool and the contents then poured into a pan of water to which a little detergent has been added. The perfectly spherical grains fall to the bottom of the pan and the charcoal may be carefully rinsed away.

The grains are free of oxidation because of the reducing atmosphere of the surrounding charcoal, and may be graded to size through fine mesh screens as used in glaze preparation in ceramics.

The fusion of the grains to each other and to more solid surfaces evidently is the result of a molecular exchange. Basically, the process consists of using a form of copper salt mixed with some sort of organic glue or adhesive. This solution is used to cement the small grains or, if desired, other small shapes, into position. The entire work is heated with a reducing flame until the copper compound forms an oxide while the glue

148

carbonizes. The carbon of the glue combines with the oxide of copper and passes off as carbon dioxide, leaving a thin molecular layer of pure copper. This film combines with molecules of gold from both the grains and the base surface to form a strong, delicate bond.

Pliny, in the 1st century A.D., recorded that the copper salt was derived from finely powdered chrysocolla, a semi-precious stone rich in copper silicate, and an animal hide glue was used as the carbonizing adhesive. Other recorders mention the use of the gem stone malachite, which contains copper carbonate.

Another method consists of collecting scales of cuprous oxide by alternately heating and quenching a sheet of copper. These flakes of cuprous *firescale* are then finely ground with a glue solution which is then used to adhere the grains in position. Though this method does result in a strong bond, the scale particles must be very finely ground and very carefully applied in solution, since the slightest overheating causes them to pit both grains and surface material.

The problem of proper heat control is most difficult to master. Long experience evidently taught the ancient craftsmen how to determine the correct heat when using a bed of glowing charcoal as the heat source.

One researcher explains the process in the following terms:

1. Mix a cupric hydrate with a glue such as seccotone.
2. Place the grains into position with this mixture.
3. Heat the unit quite slowly in a reducing atmosphere.
4. At 100° C the water has evaporated from the glue and the cupric hydrate converts to cupric oxide.

5. At a higher temperature the glue carbonizes and the cupric oxide is reduced to metallic copper, which forms a fine coating in the joints between the grains and the base surface.
6. At about 900° C this copper begins to alloy itself, partly with the underlying gold, partly with the gold from the grains, and this alloy "solders" the grains to the base surface.
7. At a somewhat higher temperature the copper from the melted alloy diffuses gradually into the gold where—to all purposes—it disappears, as the amount of the gold is about 1000 times greater than that of the copper in the alloy.

Another possible method of achieving delicate solderless fusion is currently being investigated. This process hopes to exploit the phenomenon of *diffusion,* the process of molecular interchange by contact. If given enough time, two pieces of like metal will eventually exchange molecules at a point of contact even at room temperature. If this process is accelerated by controlled heat in the correct atmosphere for a controlled period of time, a strong and precise fusion will take place.

Shot Making

For general decorative purposes, shot may easily be made by melting small scraps of metal (or measured lengths of wire if all are to have equal diameter) with a gas-air torch flame on charcoal. Once the metal has drawn up into a sphere the air should slowly be turned off, allowing only a gas flame to bathe the shot. Done this way, the shot will cool oxide-free and with a smooth surface.

149

1

3

2

John Paul Miller, an American goldsmith, has developed the ancient art of granulation to a degree seldom seen since the Etruscan gold jewelry of the 7th to 4th centuries B.C.

By combining the textural richness of granulated gold with the brilliance of enamel, Miller has given his jewelry a timeless excellence which satisfies today as it would have in ancient times. The infinite variety of shape and surface design found in sea life has been a successful source of stimulation for this master craftsman's interpretation.

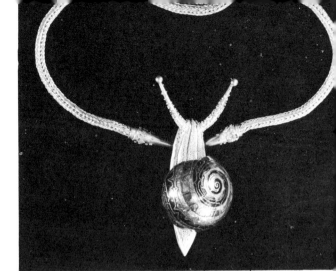

4

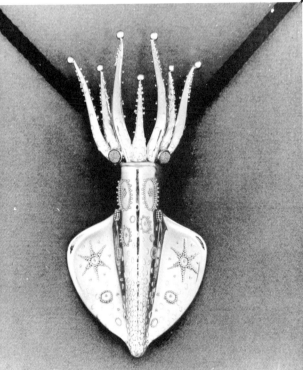

5

6

1. "Argonaut," pendant, gold and enamel; John Paul Miller.

2. "Fiddler Crab," pendant, gold and enamel; John Paul Miller.

   Photo by Frasher

3. Brooch, gold; John Paul Miller.

4. "Snail," pendant, gold and enamel; John Paul Miller.

5. "Squid," pendant, gold; John Paul Miller.

6. "Crab," pendant, gold and enamel; John Paul Miller.

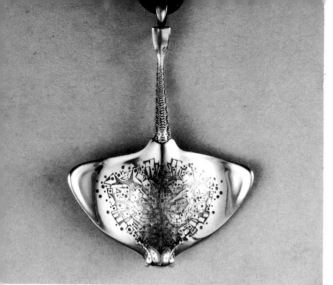

7

8

9

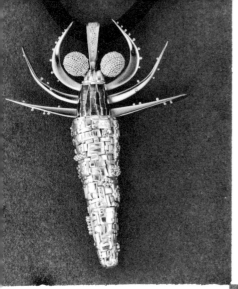

10

11

7. "Skate," pendant, gold and enamel; John Paul Miller.

8. "Cuttlefish," gold and enamel; John Paul Miller.

9. "Briars," brooch, gold; John Paul Miller.

10. "Caddis Worm," pendant, gold and enamel; John Paul Miller.

11. Marine motifs, necklace, gold; John Paul Miller.

For many years, Professor Elisabeth Treskow of Cologne, Germany, has been acknowledged as one of the few master goldsmiths this century has produced. Her work, in both design and technique, has been in the vanguard of that of a group of superior craftsmen produced by Europe in the traditions of the finest guild systems of the Renaissance.

One of the first in modern times to master the difficult process of granulation, Professor Treskow has used it successfully in the surface enrichment of her extremely personal and technically perfect jewelry.

1. Pendant in gold granulation and precious gems; Elisabeth Treskow.

2. Brooch in gold granulation, pearls, and precious gems; Elisabeth Treskow.

3. Bracelet of gold granulation, pearls, and precious gems; Elisabeth Treskow.

4. Necklace of gold granulation; Elisabeth Treskow.

5. Detail of the chain of office of the Mayor of Cologne, "The Three Kings." Gold granulation, enamel, chased gold, and gems; Elisabeth Treskow.

Photos by Elisabeth Treskow

1

2

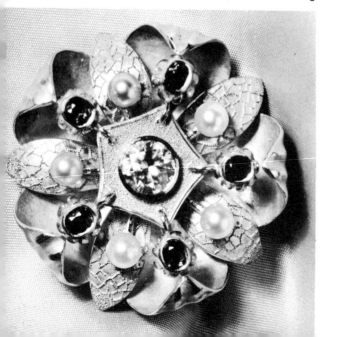

3

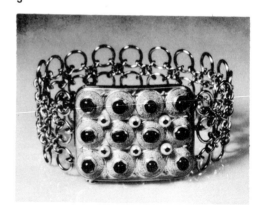

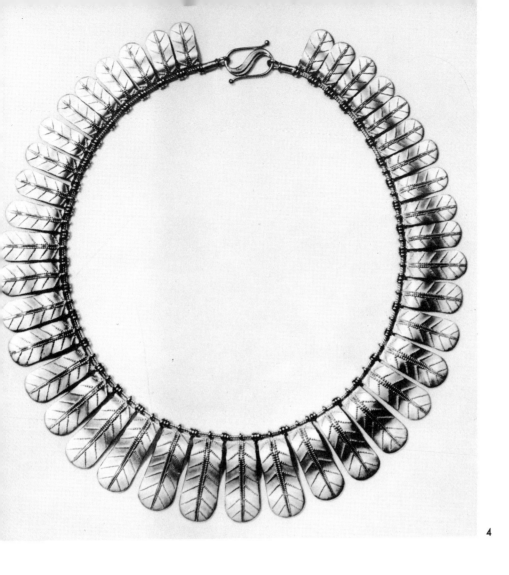

4

5

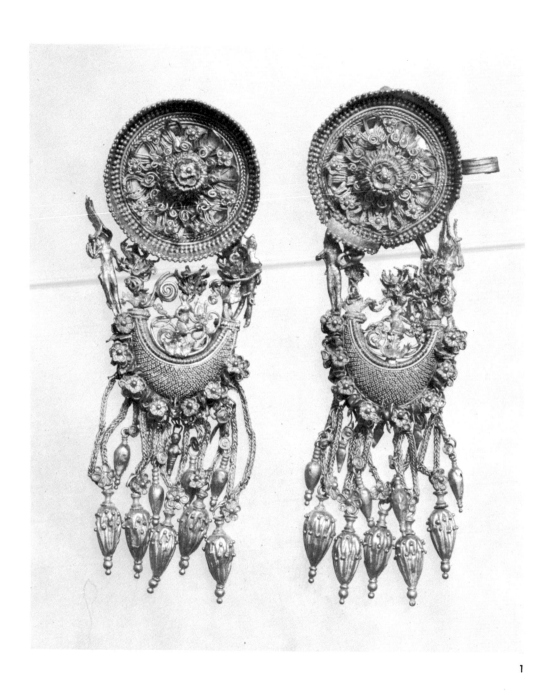

1

2

No craftsmen in metal have ever developed the delicacy of design in gold work to quite the degree achieved by the Etruscans and Greeks of the 7th to the 4th centuries B.C. Though tools and methods of working were primitive compared to those of today, they were handled with such complete knowledge and authority that it is difficult to believe that work was done without magnification or other aids.

In more recent times, from the 16th century to the present, many artist-craftsmen have tried to emulate these early masters. Some have succeeded in close approximations; most, only in clumsy efforts little related to the exquisite precision and delicacy of Greek and Etruscan work.

No records exist which give an accurate account of materials and processes used in adhering the minute, perfect spheres of gold to a surface or to each other; this has made it necessary for each later craftsman to virtually re-invent the process for himself. That two renowned artist-craftsmen have succeeded in this task can be seen in the work of Professor Elisabeth Treskow of Germany and John Paul Miller of the United States.

Both of these excellent goldsmiths have not only developed the techniques but also captured the aesthetic spirit of the past in order to develop a highly personal and contemporary expression of the art of granulation.

3

1. Gold earrings, 4th century B.C., Greek. From Madytos, Greece.
   The Metropolitan Museum of Art, Rogers Fund, 1908
2. Gold fibula (cloak pin), 7th century B.C., Etruscan. From Rusellae, Italy. The Metropolitan Museum of Art, Purchased by Subscription, 1895
3. Gold fibula, 7th century B.C., Etruscan.
   The Metropolitan Museum of Art, Fletcher Fund, 1931
4. Gold earrings or buttons, 6th century B.C., Etruscan.
   The Metropolitan Museum of Art, Rogers Fund, 1913

4

5 · SHAPES OF JEWELRY
POSSIBILITIES AND LIMITATIONS

Jewelry has many shapes, many uses. It may vary in size from a tiny accent pin to a spectacular necklace. The only real limitations to shape and size are those that the wearer or prevailing styles might impose. As jewelry contains a personality within itself, so it may ally itself with the personality of the user.

Many jewelry designer-craftsmen work directly with the client in order to evolve a design most admirably suited to the client's vision.

Another designer-craftsman may prefer to design to satisfy his own personal vision, without any specific user in mind. In this case, there might be a somewhat greater regard for the absolute limits of size and weight in the object. Since these limitations have been quite well established through centuries of use, common sense and a little insight are really all that is necessary.

Each century, and almost each society within that century, has had its most fashionable jewelry form. To the Egyptians, the large pendent collar necklace was very important. In the Renaissance, a rich linked chain and several ornate finger rings on each hand were the mark of the well-dressed aristocrat. The style of dress in many parts of India and Southeast Asia requires that a jewel be worn in the side of the nose and that

rings be worn on toes as well as fingers.

It has only been within recent times that both men and women of the West have limited jewelry uses virtually to rings, cuff links and tie ornaments for men, and to earrings, necklaces, pins and finger rings for women. Fortunately, within these rather arbitrary restrictions, there is a wealth of variety, a limitless world of ideas available to the designer-craftsman.

The design and construction details of a number of basic jewelry forms will be described in this chapter. The list may not be as complete as possible since new approaches to the use of jewelry are being found every day. The information is fundamental, and the designer-craftsman interested in creating jewelry will quickly develop his own variations and refinements.

PINS AND BROOCHES

The size of a pin would not be important except that weight soon becomes a serious factor as the size increases. Though a pin may be designed specifically to be worn only on heavy suit or coat materials, most pins must be light enough so that they do not stretch light dress materials.

As a general rule, sharp points should

be avoided. The pin is fixed to the cloth and will not give way when accidentally brushed by a hand or arm.

Pins are usually flat rather than three-dimensional. The reason is that the weight distribution of a work in the round could cause it to hang forward from the dress rather than resting against the surface. A slight contour usually works well if the pin assembly is fastened so that the piece is easily pinned to the dress.

Most contoured pins have either a flat back or a bar of flat metal to which the pin may be attached. (See Fig. 171.)

Applying pin assemblies is described **on pages 50-51.**

It should be mentioned that the back of a piece of jewelry should be treated with the same concern for excellent craftsmanship as the front. The bottom edges should be beveled even though on the top edges the design might require a precise angular treatment.

Remove all traces of oxidation from the back. This would rub off on clothing.

PENDANTS

A pendant may be no more than a simply set stone on a chain, or it might be a richly interpreted form in the center of related shapes which form the chain. Size and shape restrictions are limited to the potential wearer. The pendant may be quite delicate in construction since it is mobile enough to

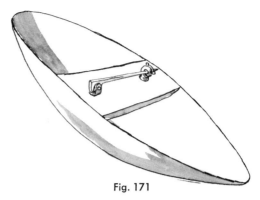

Fig. 171

move with a touch. Sharp projections should be avoided since they would catch in the dress material.

Most pendants are suspended from one point. Since they may turn during wear, it is especially important that the reverse surface be well constructed and finished. Many designer-craftsmen take advantage of this mobility by making the back as interesting as the front. Often both designs are quite different from each other, giving the pendant a double value as a decoration.

The loop through which the chain is run may be incorporated into the design of the back or the top so that its function does not distract the eye.

Simply drilling a hole at the top of the pendant through which a link is placed is too raw and insensitive for well-designed objects. The intrusion of the hole and link breaks the unity of the design and for no good purpose.

Fig. 172 illustrates a number of pendant loop possibilities.

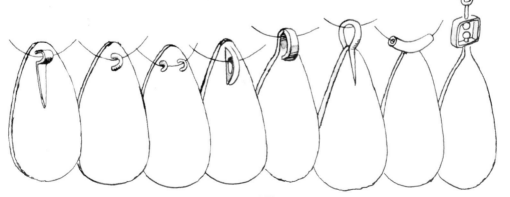

Fig. 172

1. Owl pin in silver sheet, silver wire appliqué, bezel-set opals, and ebony.
2. Single sheet of silver formed by bending and textured by chasing.
3. Bird form of sheet silver and appliquéd silver wire of small gauge.
4. Silver sheet, wire and shot with rosewood insert.
5. Sun pin of overlaid, pierced, and oxidized silver sheet.

1

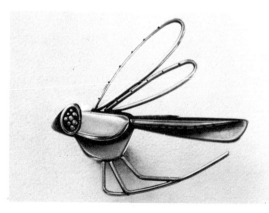

2

3

4

5

6

6. Mosquito pin of silver sheet, wire, and shot.
7. Bird pin of overlaid silver sheet.
8. *Left:* Formed and polished sheet of silver. *Right:* Seahorse pin of overlaid silver sheet.
9. Simple forms of filed and polished silver sheet.
10. *Left:* Pin of silver with copper overlay shapes. *Right:* Trumpet pin of formed silver sheet, wire, and shot.
11. Figure pin of overlaid silver sheet and wire.
12. Head of overlaid silver sheet and wire; contours formed by filing.
13. Cat pin of overlaid silver sheet, fine wire, and shot.

7

8

9

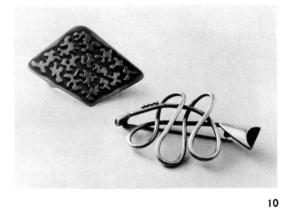

10

11

The photographs of pins on these pages are examples of University of Illinois student work and have been selected to illustrate the great variety of interpretation possible even when working with basic techniques of jewelry making.

12

13

1

2

The pendants reproduced on these two pages are the work of University of Illinois students.

3

4

1. Single sheet of silver, formed and polished.

2. Single and combined sheets of silver, formed and polished.

3. Single sheet of silver, formed and polished with matte interior section.

4. Single sheet of silver, formed and polished.

5. Single sheet of silver, formed, decorated with soldered wire appliqué and ebony set in bezel. Frame of silver inlaid with small overlay and shot; oxidized background.

6. Single sheet of silver, pierced by sawing.

7. Two sheets of silver; top sheet pierced and overlaid on solid bottom sheet.

8. Heavy gauge silver formed by filing; hammered wire inset detail.

9. Silver overlay forms on solid back; oxidation used to emphasize shapes.

10. Combined formed silver forms; pendant on right with wire applique.

9

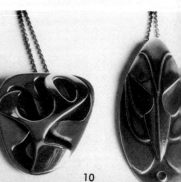

10

11

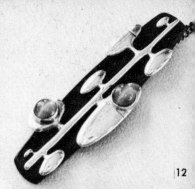

12

13

14

15

16

17

18. *Left:* Pendant of silver, ivory, and ebony; tubing separates units. *Right:* Pendant of sheet silver and carved ebony.

19. Single sheets of silver pierced and textured with chasing tools.

20. Silver wire of various gauges soldered into silver sheet.

21. Sheets of formed silver holding polished agate slab.

11. Frame of 14K gold holding suspended topaz quartz gem.

12. Heavy gauge silver, filed, set with tiger eye gems and mounted on ebony.

13. Silver overlay with wire and bezel-mounted hematite gems.

14. Formed silver wire.

15. Formed silver with tourmaline gem.

16. Upper pendant of ebony with silver and gold insets. Silver tubing forms holder for chain. Lower pendant of silver and bronze with carved ebony backing.

17. Pendants of silver made by fusing sheet by excess heating. Some detail added by chasing.

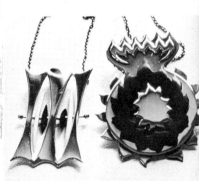

18

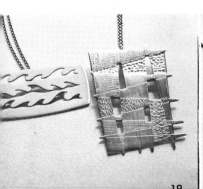

19

20

21

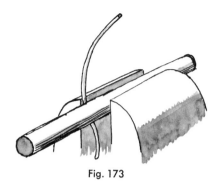

Fig. 173

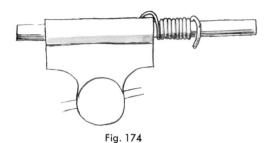

Fig. 174

Since hanging or falling natural objects are usually larger below a horizontal median (as in water drops), the eye finds it more comfortable when a pendant is smaller at the top than at the bottom. Pendants of this shape give a sense of stability. This principle could also be applied to pendants of a basic linear form, such as those made of wire or sheet-metal strips on edge.

The chains of pendants and necklaces in general average 18″ to 20″ in length for most purposes.

NECKLACES

Necklaces may be of several basic types. They may consist of identical or alternating links of wire or be made up of articulated units of sheet metal. They may be collars—solid in form but flexible enough to slip around the neck.

Necklaces may combine units of wood, plastic, bone, ivory, etc., with alternating metal forms for variety in color and surface.

LINK NECKLACES

Links have been made for at least 3,000 years—and the method is still as simple today. For round links of the same size, a rod of wood or metal just the size of the inside diameter of the desired link is wrapped with one or two turns of wax paper. This allows easy removal of links later.

Wire of the desired gauge and shape is first annealed and pickled and then wrapped in a tight spiral around the rod.

A simple way to do this is to clamp one end of the wire into a vise and alongside one end of the rod. (See Fig. 173.)

The wire is wound around the rod as tightly and as evenly as possible. Each complete turn will make one link, so it is possible to predict the number as you work. (See Fig. 174.)

The coiled wire on the rod is next placed into the jaws of the vise and all the links cut through at once with a fine jeweler's saw blade. (See Fig. 175.)

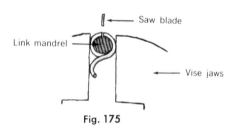

Fig. 175

Each link will now slide off as a separate but incomplete circle. Often a fine barrette file must be used to remove a burr left by sawing, since the link ends must meet perfectly for safe, efficient soldering.

The links may need flattening in addition. This may be done by bending and counterbending the ends together with two chain pliers. Avoid marring the links by using controlled pressures and by avoiding pliers with rough

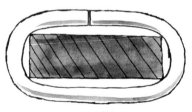

Fig. 176

164

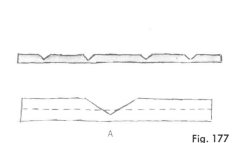
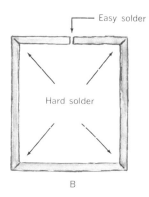

A   **Fig. 177**   B

edges and surfaces. They have no place on a jeweler's bench, anyway!

Oval links of round or square wire may be made in the same way by using a rod with a rectangular section. The mass of the wire will cause a rounded rather than an angled corner as the wire is wound around the rod. When heavy wire is used, it is best to slightly round the edges of a metal rod to prevent cutting into the wire during the winding.

Square or oblong links may be made by hammering the wire around a rod of the correct shape in order to form the angled corners. They may also be made individually by scoring the inside edge of the wire in four places. This may be done with a triangular needle file on heavy wire or by tapping a knife edge lightly into the wire if it is of a small gauge. These miters will allow the bending of a sharp corner, but may be weak. On large links or boxes made of sheet, it is best to run a small seam of *hard* solder into each bend before soldering the ends together with *easy* solder. Note that the notch in Fig. 177A is cut a little past the median line of the wire thickness. This is necessary for sharp, clean angles.

Links may be combined to make the necklace by soldering units of two together, joining two such units with a single link to make a chain of five, and so forth.

Applying the solder is simple. Have each link join clean and fluxed. Melt very small *paillons* of solder into shot

on a charcoal block. Using a small soft flame to heat the link to the soldering temperature, pick up a solder grain with a flux-moistened steel pointer and place the grain on the join at the right moment. With a little practice this may be done with great speed. Use just enough solder to fill the join since an excess causes a lump on the join which may be impossible to remove and may cause links to fuse together completely.

Solder each join as far from the next join as possible, and use a thin dab of yellow ocher to prevent remelting of joins once they are fused.

Be sure to examine each join carefully since an unsoldered link will eventually cause the chain to break.

A simple stand may be constructed to facilitate soldering of links. It is made of heavy gauge, annealed and oxidized iron wire to prevent solder adherence. (See Fig. 178.)

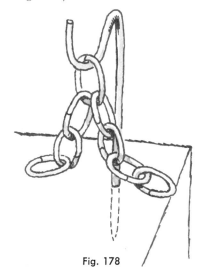

Fig. 178

165

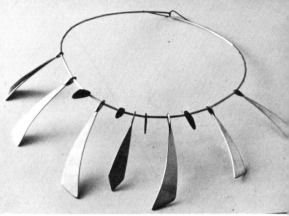

3

The necklaces photographed on this page are the work of University of Illinois students.

1. The three central units are made of overlaid silver with shot appliqué. Negative areas have been pierced by drilling and sawing. The chain is constructed of square silver wire with connecting rivets ending in shot.

2. This pendant and chain necklace is constructed of heavy gauge silver wire, tapered and ending in hollow silver spheres, and heavy gauge sheet silver filed and polished to give contour.

3. This silver collar necklace consists of a silver wire, made springy by drawing through a draw plate. Sections of small silver tubing keep the units of sheet silver and thin ebony spaced properly.

1

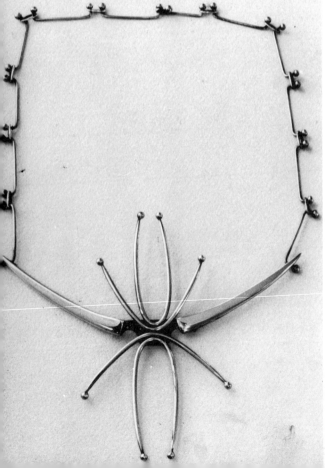

2

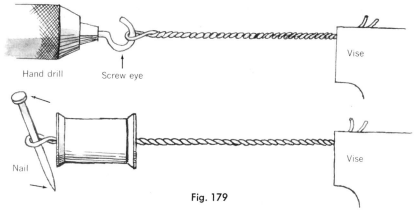

Hand drill   Screw eye   Vise

Nail   Vise

Fig. 179

There are countless link combinations which may be used to make up chains of either massive or delicate quality. References may be found in Chapter 7 to books showing diagrams and formulas for a number of interesting link chains.

Links may be made of twisted wire which allows a greater use of oxidation to contrast highly polished areas. Two or more strands of round, square, oblong, half-round, or triangular wire may be twisted together to make the wire from which the links are formed. Use of wire of different gauges and different metals often results in interesting link forms.

Wire may be twisted in several ways as in Fig. 179.

A·single strand of square or oblong wire may be twisted in the same way.

Wire should be well annealed before twisting since it quickly becomes brittle.

Unit Necklaces

These necklaces are made up of individual shapes of metal or other materials, joined together by some sort of linkage for flexibility. The linkage may form an important aspect of the total design or it may be kept hidden and minimized. Units may be strung together on a small but sturdy machine-made chain if the contrast is not too distracting.

The following are a number of unit-linkage possibilities. Fig. 180A uses half-links soldered to the back of each unit. A fine chain joins the units and hollow spheres are used to maintain the spacing. Without spacers of some sort the

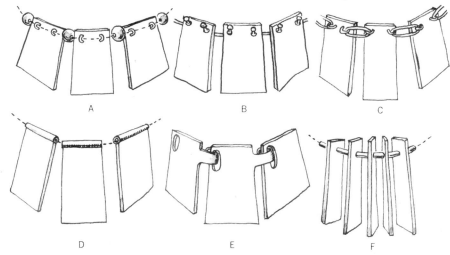

A   B   C

D   E   F

Fig. 180

1

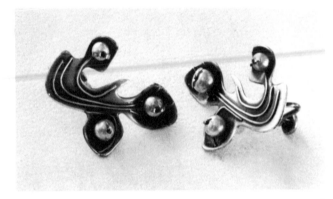

2

The earrings photographed on this page are the work of University of Illinois students.

1. *Left:* Button earrings of silver sheet with inlays of ebony and ivory. Silver strips hanging below swing freely during wear. *Right:* Button earrings of chased and overlaid silver sheet with bezel-mounted tiger eye gems.

2. Button earrings of formed silver with appliqué wire and peg-mounted pearls.

3. *Upper Left:* Button earrings of silver sheet and wire with bezel-mounted garnets. *Lower Center:* Pendant earrings of formed silver sheet. Rib shapes move when earrings are worn. *Upper Right:* Pendant earrings of formed silver sheet. Units surrounding bezel-mounted onyx gems rotate when earrings are worn.

4. Button earrings of silver sheet and wire with bezel-mounted onyx gems.

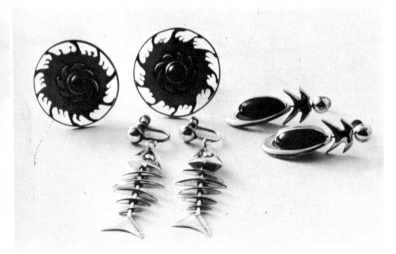

3

4

168

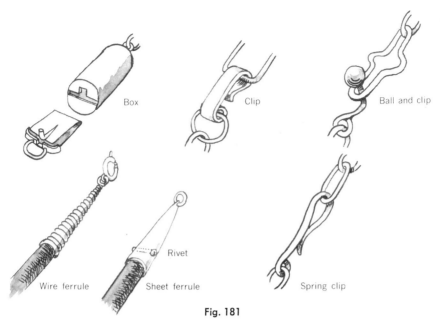

Box

Clip

Ball and clip

Rivet

Wire ferrule

Sheet ferrule

Spring clip

Fig. 181

weight of the units will cause them to slide together during wear.

Fig. 180B uses sections of U-shaped wire through drilled holes in each unit and headed with a grain of shot.

Fig. 180C uses horizontal half-links to connect a long oval link between each unit.

Fig. 180D uses sections of seamless tubing through which a fine chain is run. The tubing could be placed below the top of the unit as well.

Fig. 180E uses a tongue projecting from the edge of each unit which is bent through and around a hole or slot in the next unit.

Fig. 180F uses loose sections of tubing as spacers for edged strips of metal, wood, etc., and also to hide the machine-made chain.

Thongs and cords of leather and cloth fiber are often used as chains or suspensions, but these wear out and easily become soiled. When using such materials, design for occasional and easy replacement.

Necklace clasps may be machine-made or, better, designed as a part of the necklace itself.

The diagrams in Fig. 181 illustrate a few of the basic fastenings. Variations and new forms are constantly being developed by imaginative jewelry designers.

EARRINGS

Basically of two types, button and pendent, earrings offer an excellent opportunity for light and delicate handling of precious materials. They must be light in weight to be comfortable, and since they are worn in a safe zone, thin wire and thin sheets of metal may be used with greater assurance than in other jewelry forms.

Though it is possible to construct the earring backs, it is often a wasted effort since the best manufactured findings are both sturdy and pleasing in design. Above all they work efficiently. Earring findings are of four types: screw and patch, clips, pierced-ear wires, and pierced-ear screws.

The directions for applying earring findings may be found on page 52.

Again, the size of an earring can be a personal choice. On the average, an over-all diameter of 1″ for button-type earrings—those worn at the earlobe—is customary. For pendent earrings—those which are mobile and hang below the

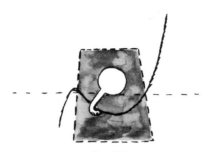

Fig. 182

earlobe—a length of 2″ to 3″ may be used.

The correct location of the clip or screw wire is important since an incorrect position could cause the earring to be worn askew or to hang forward, bending the earlobe inward with its weight.

Fig. 182 diagrams the correct position of a backing for a button-type earring. The patch of the earring back should be soft soldered above the median line of the earring so that the greater weight hangs below the finding. The ear back should be angled so that the wire curve fits the lobe near the junction with the neck.

A flat treatment of single sheets or overlays of metals 20 gauge B and S or thinner seldom results in a heavy earring. For more three-dimensional shapes, a forming or repoussé technique in thin gauges would be appropriate.

Pendent earrings are usually more complex than the foregoing since an element of mobility must be considered. These earrings are of two parts: the fastening, to which the ear back is soldered, and the pendent form itself. Some craftsmen use a manufactured fastening which is really a part of the ear back.

These ear backs come with a small ring built onto the curved wire from which the pendant is suspended. Some types have bosses—half domes—fixed to the back of the patch itself as a decoration. Unfortunately, these arbitrary machine-produced findings seldom integrate well with the design of the earring itself. It is much better to use a standard ear back and to solder it to a designed form to which the pendant is linked.

For complete mobility, fastening and pendant must be joined in a way that prevents pinching of links or earlobe. Fig. 183 shows a basic combination.

Wire, by itself or combined with sheet, is an ideal material for pendent earrings.

For pierced ears, not as uncommon as one might believe, both button and pendent forms are normally used.

Findings for button types consist of a fine-threaded rod soldered to a small patch which, in turn, is soft-soldered to the back of the earring. The rod is inserted through the hole in the earlobe and held in position by a small screw cap (Fig. 184A). A simple form consists merely of a small wire rod to which a spring clip is attached (Fig. 184B).

Pendent earrings are also suspended directly from thin wire loops which op-

Fig. 183

erate on the safety-pin principle. (See Fig. 185.)

Some people are sensitive to copper in any form, so pierced ear wires are often made of high karat gold wire.

There has been considerable interest in developing new forms of ear decoration. Jewelry has been designed to be worn at various points on the outside of the ear, surrounding the entire ear, or at points within the ear. The problems of developing safe, comfortable, and unobtrusive findings for these new positions are, of course, greater, but the impact of new uses such as these is often fresh and delightful—as jewelry should be.

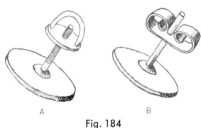

A   B

Fig. 184

BRACELETS

After neck ornaments, bracelets for the arm, wrist, or ankle are, perhaps, the oldest forms of jewelry. Though it is not now the custom in the western world to wear arm and ankle bracelets, this direction should be explored as a fertile field for innovation.

Wrist bracelets, like necklaces, may be made of one piece, to be clipped on or slid over the hand, or they may be articulated with links or units.

Solid clip-on bracelets are usually made of a thick gauge of metal—12 to 14 gauge—so that the necessary springiness may be maintained. If thinner gauges are used, the constant working of the metal could eventually cause brittleness and cracking. If thin gauges are necessary, it is best to use an overlaid design to re-enforce the bracelets where the curve is greatest. If nonmetallic decorations are used, they should be placed where the clipping action will not spring them from their settings.

Clip-on bracelets are difficult to design for general use—each bracelet should

Fig. 185

be designed for the individual wearing it, if possible. Where this is impossible, a length of 6½″ to 7″ leaving an opening of 1″ to 1½″ for the wrist to enter is adequate. The ends of the bracelet should always be rounded so that it may be fitted onto the wrist painlessly.

It is a matter of choice whether to solder overlay or other fittings to the bracelet before or after curving it to form.

If soldered before bending, the overlay joins must be absolutely sound or the tension of bending will surely cause parts to lift or break off.

Where delicate forms are to be applied, soldering must be done after bending, since bending, in most cases, involves considerable hammering with wood or plastic mallets. This hammering could distort or break delicate wire or overlaid forms. Repoussé designs must also be applied after bending since the raised areas would distort or stretch to breaking if bent too much.

The advantage in bending the bracelet to shape as a final step is that, since all soldering has been completed and no additional heat softening will occur, the bending itself will cause considerable springiness to develop.

When units are to be soldered to the curved bracelet, they must each be pre-bent or filed to conform perfectly to the surface to which they are soldered.

There are large oval bracelet mandrels available which are a help in bending not only the bracelet but also the added forms. Where the bracelet is narrow or constructed of heavy wire, it may most safely be bent to shape by hand. For comfort it is important to bevel at least all of the inner edges, as a matter of course.

Linked bracelets present many of the same construction possibilities and problems described in the section on unit necklaces. Linkages and fastenings are also similar, except that much may be done by using tubing for hinges in bracelets. (See Fig. 186.)

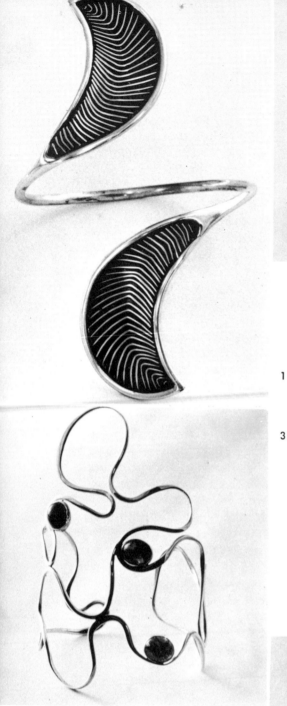

1

3

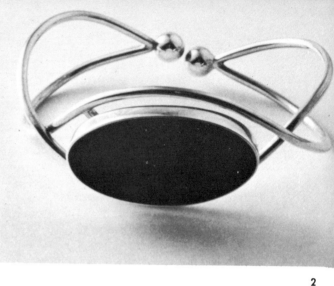

2

The bracelets on this page are the work of University of Illinois students.

1. Clip-on bracelet of silver sheet, wire, and fine wire appliqué.

2. Clip-on bracelet of heavy-gauge silver wire, hollow spheres, and bezel-mounted ebony.

3. Clip-on bracelet of heavy-gauge silver wire, contoured by hammering and set with moss agates.

4. Unit bracelet of linked forms of silver sheet and wire overlay and pierced and solid forms.

5. Clip-on bracelet of silver sheet with sheet, wire, and shot applique.

5

4

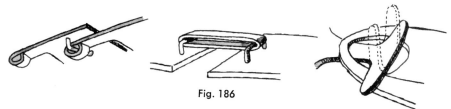

Fig. 186

A hinge may be hidden on the inside of the bracelet or it may be visible between the units.

One of the many ways in which hinges may be constructed is as follows:

1. File the edge of the unit to conform with the side of the tubing (Fig. 187A).
2. Cut a section of tubing $\frac{1}{16}''$ longer than the edge of the unit, and again cut it into three equal parts.
3. Solder the two outside sections of tubing to the edge of one unit, allowing the tube ends to be level with the outside edge of the unit. Solder the center section of tubing to the center of the edge of the opposing unit (Fig. 187B).
4. Check the fit and carefully file away just enough to form a tight fitting between each of the three sections of tubing. There should be about $\frac{1}{32}''$ to file away on the inner ends of each of the outside sections.

5. Use a silver or gold wire hinge pin just the diameter of the inside of the tubing. This pin should be $\frac{1}{16}''$ longer than the combined sections of tubing and should have a disc just the size of the tubing and of 24 gauge metal soldered to one end (Fig. 187C).
6. Connect the hinges by inserting the pin through all three tubes and carefully hammering down on the projecting $\frac{1}{16}''$ to form a rivet button (Fig. 187D).

RINGS

Though they are often simple enough to construct, rings present difficulties which do not occur in other forms of jewelry. The main problem is that they must fit comfortably.

Ring sizes range on a scale from No. 0 to No. 13½. Each size differs by 0.032″ from the next full size in diameter. Mandrels are made that are ringed with numbered grooves, each indicating a

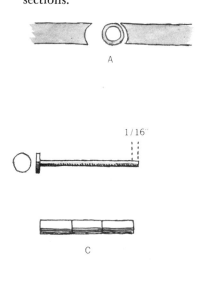

A

1/16″

C

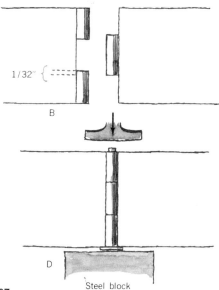

1/32″

B

D

Steel block

Fig. 187

1

2

The rings on this page are the work of University of Illinois students.

1. Cast sterling silver ring holding a pearl with two pegs.
2. Cast 14K yellow gold ring with alexandrite gem.
3. Sterling silver ring of formed sheet stock with tiger eye gem.
4. Sterling silver ring of formed sheet stock, soldered overlay and chased texture, black onyx gem.
5. Sterling silver ring of three bands of heavy gauge sheet.

3

4

5

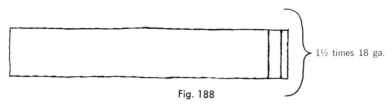

1½ times 18 ga.

Fig. 188

ring size. A set of graduated ring sizes, each stamped with the number of a full or half size, is useful in determining the diameter of a ring.

A simple band of the correct diameter and shape is often the base for more complex ring designs. The following describes the construction of a simple band or *shank*.

A Simple Ring Band (Shank)

The gauge of sheet metal or round or rectangular wire for a ring band must be determined by the intended use for the ring (occasional or constant), or the preference of the wearer. Where heavy duty durability is not most important, appearance becomes the determining factor. In general, a man's ring is two to four gauges thicker than a woman's ring: 18 gauge B and S for a woman's ring and 14 to 16 gauge for a man's ring are average gauge sizes.

The width of the band can be determined only by design principles. Comfort in wearing must be kept in mind while designing a ring. A band that is too broad can cause skin irritations due to collection of moisture and dirt.

Ring Band Construction

1. Measuring the blank. If you know the ring size in inches, add 1½ times the gauge or thickness of the metal to insure sufficient length. *Example:*

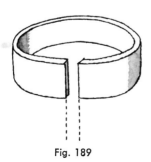

Fig. 189

The circumference of the finger is 2"; the metal is 18 gauge; then the total length would be as shown in Fig. 188.

2. Circumference can be measured quite accurately by cutting a ¼" x 3" strip of heavy paper and curving it around the large *knuckle* of the wearing finger. Make certain that the paper is curved around loosely enough to indicate accurate length. Mark the paper at the overlapping point. Add 1½ times the gauge of the metal and transfer these dimensions to the metal.

3. Some mandrels are made with a scale on the handle showing ring sizes in inches. Again, be sure to add 1½ times the metal gauge to the total length.

4. After sawing out the shape, do not file more than the ends of the blank. These *must* be true and parallel for a sound solder join.

5. Bend the ring ends together by hand, by ring-forming pliers, or with a plastic mallet. Do not be concerned with roundness or symmetry at this point. (See Fig. 189.)

6. By bending the ends back and forth and over and under each other, a spring or press fit develops to hold the ends firmly together. It is not necessary to use binding wire if this has been carefully done. Check to make certain that the ends meet cleanly for the entire length of the join. (See Fig. 190.)

7. Prepare the band by propping it on charcoal or pumice pebbles or by holding it in spring-locking soldering tweezers. The join should face up and should be covered by one piece of *hard* or *I T* solder next to

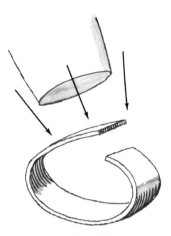

Fig. 190

the other along its entire length. It is a good plan to use a bit more solder than usual and to allow the excess to remain until all other soldering has been completed. This helps to prevent the disintegration of the solder seam during future soldering. (See Fig. 191.)

8. Heat the ring so that both sides of the join absorb equal heat. Unequal heating could attract the melting solder away from the join—it must be carefully avoided. Heat until the solder has been fused down into the join; check this by examining the under side. Then quench the band in cold pickle or water as quickly as possible.

9. Remove all traces of flux *glass* with hot water. Once again examine the join and add more solder if necessary.

10. If the join is solid, the ring may be *trued* or made round on the ring mandrel. Using the mandrel as a support, shape the ring into circular form with a wood mallet against the support of the bench. (See Fig. 192.)

11. The final truing takes place while hitting the edge of the band lightly with the edge of the planishing hammer. Since the mandrel is tapered, it is important to reverse the ring several times during this process (see Fig. 193). If the ring is too small, it can usually be enlarged by

striking the surface of the ring with the face of the planishing hammer. This tends to thin the gauge of the ring rather quickly, so care should be exercised during this step. If the band is too large after truing, it must have a section cut out (on both sides of the first solder join) and then may be resoldered as before.

12. Once the band fits well, the necessary filing of inside and outside edges and surfaces can take place. If nothing is to be added, the band may be buffed and polished by hand or, supported safely with a mandrel, with the buffing machine. There are several types of buffing mandrels designed for the inside surfaces of a ring.

13. After the band is thoroughly finished, other parts such as tables, wire applique, bezels, etc., may be added, using lower melting solders such as *medium* and *easy*.

The basic band may be pierced for positive and negative shapes before bending or it may be filed on edges and surfaces to bring out relief areas.

Always solder tables, etc., over the band solder join. For a description of basic bezel construction, see the section on gems in Chapter 4.

Variations of the basic ring shank may be combined with tables or other forms in the following ways (see Fig. 194).

Ring bands may be made of wire using various combinations of shapes and gauges. The measurement plan is the same as for sheet. The wires may be soldered together before bending by placing the accurately measured pieces

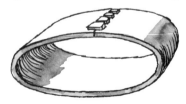

Fig. 191

carefully side by side. Pieces of *hard* solder, enough to fill the entire join between each wire, are placed at one or both ends of the combined wires. If the wire ring is to be formed into a circle, the ends should be filed to angles to make a smooth, strong join. When soldering with *hard* solder there are often small unmelted lumps left after the join has been filled. These may be removed with a fine triangle or barrette needle file and an edged scotch stone.

If two or more wires are to form the base support for a table, they may be soldered quite easily in the following manner:

1. Construct and solder the individual wires so that they are identical in shape and diameter (Fig. 197A).
2. If three rings are used, sand a slight flatness on the side of ring No. 1 by rubbing it on a flat sheet of medium emery paper. After sanding, the edge should touch all around the circumference of the ring when placed on a level steel block.
3. Flatten ring No. 2 on both edges in the same way.
4. Flatten ring No. 3 only on one side.
5. Place No. 1, flat edge up, on a smooth charcoal block. Paint one half of the circumference with flux and the other half with yellow ocher. Place two or more pieces of *hard* solder at a point farthest away from the first ring join and in the center of the fluxed area.

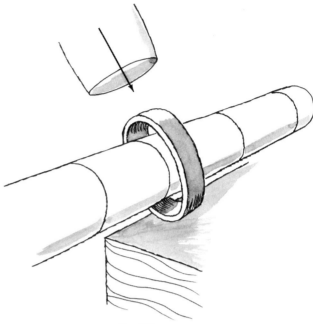

Fig. 192

6. Place ring No. 2, fluxed and painted with yellow ocher, in the same way as ring No. 1, on top of the first ring. Add solder to No. 2 and place No. 3, also painted with ocher and flux, on top of No. 2.
7. Solder the rings together by concentrating the heat on the fluxed side of the rings. The yellow ocher should prevent the solder from flowing completely around the rings (Fig. 197B).
8. After pickling and washing, the rings may be pried apart in their unsoldered sections by inserting a knife blade and carefully twisting to the desired width (Fig. 197C).
9. The points to which a table may be soldered may be filed level and the

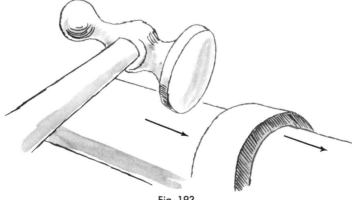

Fig. 193

177

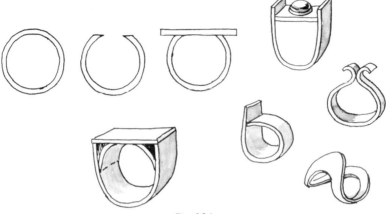

Fig. 194

band soldered to the table from behind with *easy* or *medium* solder (Fig. 197D).

Another basic ring-forming technique forms the band and the table base at the same time, as in Fig. 198.

Variations of the above shape are suggested by Fig. 199.

It is difficult to pre-size a ring of this design so it often helps to make a model of heavy paper, using the final paper forms as templates for shaping the metal.

A final suggestion on ring forms consists of cutting the entire ring shape out of two or more sheets of metal and, as in the above design, soldering only half of the shank together. (See Fig. 200.)

CUFF LINKS AND STUDS

In cuff links the designer finds the only real latitude for fresh design invention in men's jewelry. Though size is limited somewhat by usage and function, the weight and individuality of a design offer free rein.

Cuff links may be constructed of sheet metal, may be formed, may use wire ap-

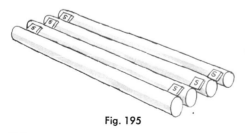

Fig. 195

plique or free-standing wire shapes, or may be cast. The only real considerations are that they fit well and that they be safe to wear. Many fine designs would not be practical because of sharp projections or thin wire additions. The average cuff link size is a square or circle of 1″ diameter though the designer has considerable leeway in this.

Since the design and construction problems of the face of the cuff link are the same as those for pins, pendants, and

Fig. 196

earrings, it should be necessary only to describe the various fastening possibilities.

On commercial swivel-bar cuff links, several of which are quite cleanly designed, the distance between the swivel bar and the cuff link face is about $\frac{9}{16}″$, though this could be somewhat shorter for women's cuff links.

There are two types of commercial link findings suitable to fine jewelry. One type has a soft solder patch for use with enamels, and the other, really the better, has two parts (Fig. 201).

1

2

The cuff links on this page are the work of University of Illinois students.

1. *Left:* Sterling silver cuff links of sheet overlay and wire and shot appliqué. *Right:* Cuff links of ebony and cocobolo woods held in sterling silver.

2. *Left:* Cuff links with sterling silver sheet, wire, and shot appliqué. *Right:* Cuff links of fused silver with chased detail.

3. Cuff links of ebony between two sheets of sterling silver.

4. *Left:* Cuff links of formed silver with ebony insert. *Upper Right:* Sterling silver base, sides, and top. Interior of crushed turquoise and red sealing wax, filed flush with upper silver plane. *Lower Right:* Flat and formed sterling silver sheet with vertical wire detail.

5. *Upper Left:* Cuff links of raised sterling silver sheet. *Lower Left:* Cuff links of sterling silver sheet; top plane angled to form depression for sections of vertical gold wire. *Upper Right:* Cuff links of flat and formed sterling silver sheet. *Lower Right:* Cuff links of overlaid sterling silver sheet with wire bridges.

3

4

5

179

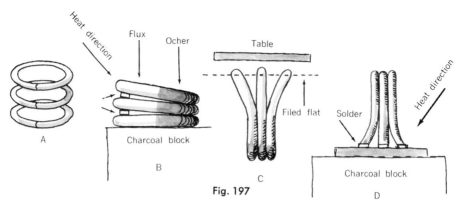

Fig. 197

Having a separate joint to which the finding is finally riveted allows the use of silver solder. Excess heat on the finding itself would quickly destroy the temper of the spring in the swivel bar.

Cuff link findings are often constructed by the craftsman and may be variations of the following (see Fig. 202).

As in earrings, it is often practical to saw and file both cuff links at the same time by holding two pieces of sheet in a vise or a ring clamp. Remember to oppose the pieces eventually so that the design means the same thing on each cuff. (See Fig. 203.)

Also as in earrings, it is not always necessary to make each half of a pair identical. A meaningful variation on a theme might be far more interesting than absolute uniformity.

BUCKLES

Buckles, like rings, present the added difficulty of having to fulfill a specific function. They must be simple enough for easy wear, sturdy enough to withstand constant use, and interesting enough to warrant substituting them for perfectly usable, if uninspired, manufactured articles. Fig. 204 indicates a few of the buckle fastenings.

In Fig. 204A the bar to which one end of the belt is sewed may be made of

Fig. 199

8 to 10 gauge wire. The curved peg is of the same gauge but should be rounded and smoothed. The loose end of the belt slides behind the buckle, the peg enters a hole, and the end is slipped into an inside leather or metal loop.

Fig. 204B shows the structure of the

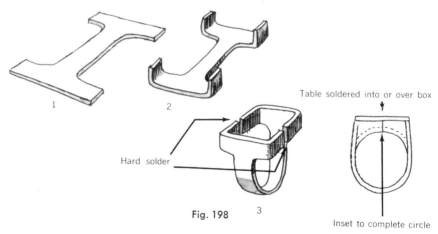

Fig. 198

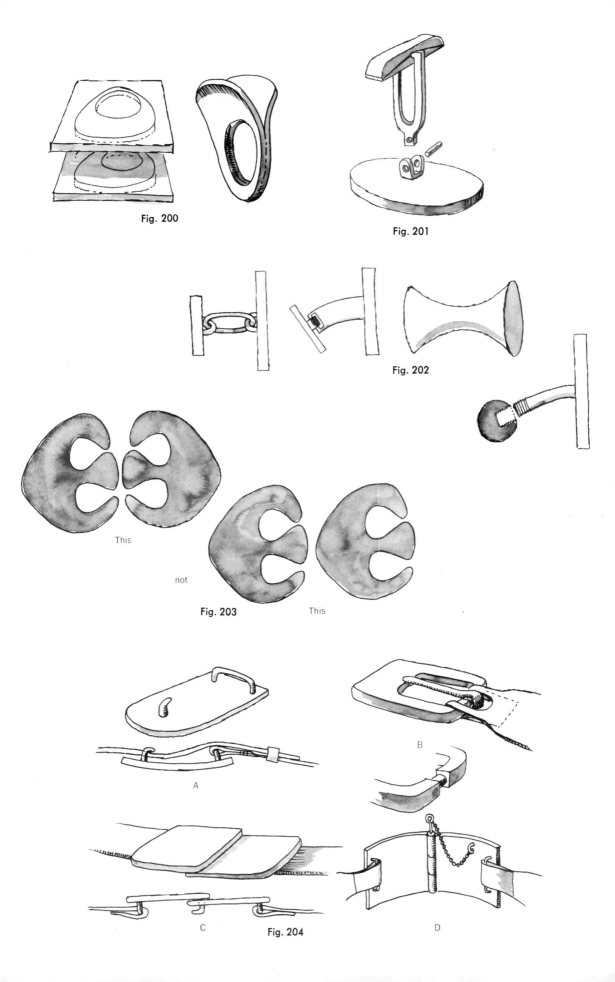

Fig. 200

Fig. 201

Fig. 202

This

not

Fig. 203

This

A

B

C

Fig. 204

D

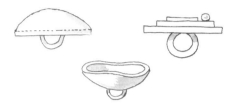

Fig. 205

conventional buckle which may, of course, be varied in proportion and size to any degree desired. The point at which the buckle tongue is bent around the buckle bar should be filed round and notched to prevent slipping of the tongue.

Fig. 204C has the two belt ends sewed to 10 or 12 gauge wire loops on the back of each buckle half. The halves are held together by one or more heavy wire hooks placed through corresponding holes. If belts of this sort must be adjustable, one belt end might be fastened to a number of snaps for this purpose.

In Fig. 204D a variation of the last buckle described uses tubing for a hinge. This tubing should have a thick wall to withstand the constant opening and closing of the buckle. The removable pin has a ring attached to one end from which a small sturdy chain runs to a loop on the inside of one buckle half. This secures the pin even when the belt is open.

If a buckle is to be made of one thickness of metal, it should be at least 12 gauge to prevent a flimsy appearance. Thickness may, of course, be built up by layers of overlaid metal, wood, or other materials.

BUTTONS

Since sewing on a button through holes in its surface could destroy much of its design unity, it is better to fasten them to the cloth from behind by sewing through rings or half-loops soldered to the back of each button. To avoid an unpleasant thinness, the button should be domed and backed with a flat sheet or built up to 8 to 10 gauge thickness.

TIE BARS AND TIE TACKS

There are two general qualifications for tie bars. They must clip well and they cannot be too heavy. Machine-made findings for tie bars, though they may function well, are of unpleasant shape and design. They seldom integrate well with the bar itself.

Tie bar clips may be made in two basic ways. The clip may be an extension of the bar, bent to shape after all soldering and oxidation has been completed. If overlay forms of sheet or wire are used, they should not be soldered closer than ⅜″ from the point of the beginning curve (Fig. 206B). It would be difficult to bend a double thickness at that point without putting a great strain on the solder join.

To maintain as much hardness as possible in the clip—a strong spring is absolutely necessary—all soldering should be done without quenching the work in cold water or pickle. Oxides may be removed by boiling pickle when the work has reached room temperature after soldering.

Before bending, the clip may be carefully planished with a smooth flat-faced hammer on a smooth flat stake. The compression of the hammer blows will increase the springiness of the metal.

The second basic construction consists of using two parts: the bar and a sep-

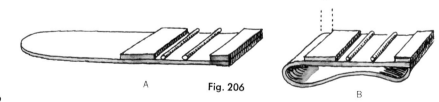

A   Fig. 206   B

1

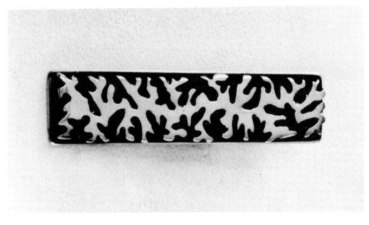

2

The jewelry on the page is the work of University of Illinois students.

1. Cuff link and tie bar set of sterling silver sheet with a brushed finish. The tie bar is of one piece, bent around to form the clip.

2. Tie bar of sterling silver sheet. Upper plane elevated on several shot of identical size. Clip of separate piece of silver sheet.

3. Hair ornament of silver sheet with ebony and ivory pin decorations.

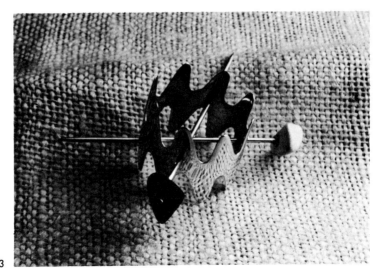

3

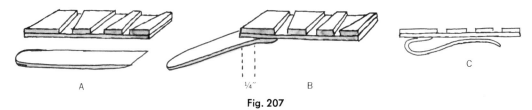

Fig. 207

arate clip form. In Fig. 207A, note that the clip section is as long as the bar itself and that the bottom edge of the squared end is sharply beveled.

In Fig. 207B the clip has been given a sharp bend so that it may be soldered to one end of the bar without having the solder run too far. The bend in the clip should be hidden. Do the bending after all soldering has been completed.

If special hard or half-hard spring sterling silver is used, the clip must be soldered on with lead solder to avoid annealing it.

If standard sterling silver is used, the same pickling procedure as described for Fig. 206 should be followed.

Bending the clip may often be done by hand, using a smooth round rod of the proper size to form the curve around. A pair of smooth-jawed bending pliers often helps to work the clip up against the back of the bar. A few careful taps—this *could* distort the bar— with a wooden mallet often help to bend

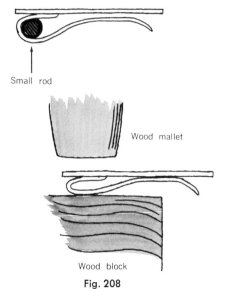

Small rod

Wood mallet

Wood block

Fig. 208

the clip against the bar. The surfaces where the bar and the clip touch may be scored cleanly with a file to supply a better grip. (See Fig. 208.)

Most ties are 2″ to 2¼″ wide, so the bar itself should be about that length.

Fig. 209

The width depends on the design, and total weight should be considered since a heavy tie bar usually works itself off.

Tie tacks use a pointed peg which is pressed through both tie and shirt to be fastened from inside the shirt with a clutch catch.

The manufactured findings are usually simple in design and very practical.

The peg is soldered to the base of the design with soft solder and should be located somewhat above the median line of the design for good balance. (See Fig. 209.)

Tie tacks may be made by soldering a pointed peg to the design back and constructing a one- or two-piece clutch catch. If possible, spring silver should be used in constructing tie tack catches.

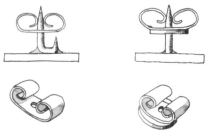

Fig. 210

6 · STIMULANTS FOR THE MIND'S EYE

To instruct in design is, in a sense, a presumption. Nothing is as personal as an individual's design idiom, and to write about the best way of developing a design idiom can result in gross generalities.

Many authors have attempted manuals of design—formulas to successful solutions of the infinite problems of design. Most have failed because they depended too heavily on the generalities and gave too little emphasis to design as a personal development.

Certainly the foundations of design—such as balance, rhythm, and variety—are sound enough, but without suggesting the wealth of visual excitement around us with which to build on these bases, no more than conformity of thinking and style can result.

In the history of art there have been centuries when a style reigned with such absolute uniformity that few artists could conceive of breaking the rules. In many cultural periods the minor changes, the subtle abstractions of natural form, had developed so slowly that eventually the idea source was lost completely and each generation continued in the tradition of the previous generation.

Today, those who live in the highly industrialized countries find the opposite to be the case. Tradition has been rather completely replaced by the concept of change for the sake of change. The old ways, simply because they are old ways, are discarded in favor of newness and planned obsolescence—the sources of a happy and successful life. This results in an ever more frantic search for a fresh approach, whether superficial or serious.

Never in the history of art has such pressure been placed on the individual artist-designer. In place of the comforting and absolute knowledge that he *exists* within his social cell, today's visual artist may search alone in a landscape without roads and without signposts.

There is a tragic anachronism in this. In museums, galleries, and libraries; on radio, in film and in television, we have virtually all of the knowledge of the past, the present, and even some of the future at our fingertips. We have an encyclopedic collection of all that man has thought and done. At the same time we move in a traditionless world, or—more exactly—in a world where we can adopt a hundred traditions as our own if we wish. Too many choices can be just as binding to the free imagination as too few.

The artist-designer must, therefore, come to an agreement within himself.

185

He must become his own critic in all things since his society can no longer accurately judge him within the framework of a tradition.

Styles and modes of expression change today much faster than ever before. In the 20th century each decade—almost each year—has had its fashionable idiom of expression. It would seem impossible for a serious and painstaking artist to anticipate each wave and consistently ride its crest. To do this is not only impossible but also invalid since a vogue may or may not have lasting value until it is viewed in distance and time.

The concept of freedom of expression is sound and healthy. The question to be answered is: freedom from what? What unbearable restrictions, what tasteless dogma? Since all is permissible, is there anything to rebel against? Many would say yes. Aesthetic conformity should breed rebellion. Within a small tribe with limited knowledge of other people and their ways, a design conformity was an understandable development. To conform to a successful style of expression today is irrational and superficial since we do know what others have done and are doing, and we can profit by the thousands of mistakes and successes of the past.

Stated simply, the artist-designer of today must be omnivorous. He must have great curiosity and, above all, he must be ever critically aware. He must be as objective about an artifact made 3,000 years ago—analyzing its value as an object of creative unity—as he is of the work of his contemporaries. He might even be a little cynical and suspicious of the success of a current mode until he gleans from it that which is universal, exploratory, or stimulating.

Since the artist may strike off in any direction he wishes, he must equip himself with knowledge. It is never enough merely to know how to manipulate materials—*that* sort of knowledge may be taught to anyone. What cannot be taught—and this separates the serious artist from the imitator—is the discovery of stimulation and excitement in form in all its variety, natural, man-made, or accidental.

This the artist-designer must do for himself. He must be an art historian, not to catalogue and describe, but rather to dissect, to analyze, and to explore. He must be a naturalist, discovering for himself the relationships of shape, color, and texture in the natural forms around him. All he sees, all he senses is stored for future use. The greater the experience of seeing and feeling, the less possible it is for the artist to be caught with only a conventional or trite solution to a design problem.

The artist must, above all, be an inventor. He must invent his personal analysis of a shape, a color, and a texture. To invent he must experiment. Each experiment should lead to a conclusion, since to experiment merely for the sake of experimentation is like building on quicksand—nothing solid can result.

SOME BASIC DESIGN PRINCIPLES

Design, whether it be for jewelry, sculpture, painting, or any of the many visual art forms, is based on simple universal principles. Though terms of definition may vary, each designer is ultimately concerned with shape, form, surface, and color.

The results of the successful use of these basics may express tension or repose, stability or flux, the minute versus the mammoth, the rhythmic against the unpredictable. Each of these almost emotional responses may be emphasized or understated to make an individual work of art.

The term *design* has, for many, the meaning of balance— even bisymmetrical balance. Though world art is full of many handsome examples of bisymmetric balance, it is only one way of arriving at this all-important element in a work of art. In Fig. 211A, bisymmetric balance is shown as an arbitrary arrangement of shapes. **X** is exactly balanced by **Y** with **Z** in the exact center.

The emotional response to a design of this sort is one of security, lack of surprise, and—after a short while—apathy. The eye quickly recognizes the interrelationships of the shapes and finds no more that is stimulating.

Fig. 211B has also arrived at balance —a dynamic balance, a balance of constant movement over basic stability. The large form—**Z**—though placed to one side of the center, does not overbalance the rectangle since the **X** and **Y** forms flow between the halves and, because of a multitude of angles, create enough stimulation to balance the massive attraction of the large triangle.

In addition, the relationship between each small form and its neighbor is different from every viewpoint. There is no easy prediction of spaces, and so—no quick apathy. This brings the problem of containment into position.

All visual art forms, with perhaps the exception of mobile sculpture, exists within finite boundaries. A painting lives in its rectangle, a piece of sculpture within its envelope of material, a piece of jewelry within the practical size and shape it must have.

Many sculptural forms send out lines of motion which are not contained by a frame. The rooms or landscapes in which they rest form the limits of shape and form just as well.

The idea is that the eye should be brought back to the subject rather than be led constantly and irrevocably out and away. For centuries painters have worked on methods of composing so that the eye remains within the picture, finding all that is necessary there.

In the design of a piece of jewelry it is often helpful to establish an arbitrary limitation. This could, logically, be the maximum size of the object as defined by a circle, an oval, a square, or a rectangle.

Whether the design is an abstraction of a natural form or a nonobjective invention, the discipline of altering shape to a containing limitation often helps to develop unity and compactness.

The concept of compactness in design has its historic basis. Perhaps the most extreme, and successful, example of this is seen in Scytho-Sarmatian art. These people, perhaps as an involuntary reaction to the limitless space of the Asian steppes from which they came, had a strong sense of *horror vacui*—they felt that uninvolved space was uncomfortable in their art. Consequently, the design of the culture is known for its compact use of shape. Forms, usually animal subjects, were often distorted to fit compactly

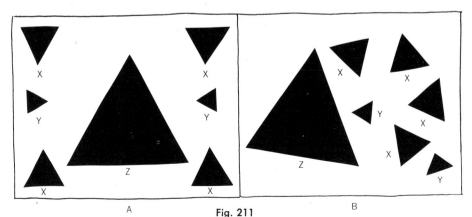

A   Fig. 211   B

1

The most striking characteristic of the work of Robert Pierron is the precision of his abstract sense and the clean
craftsmanship of each piece of work.

Whether the subject be human, animal, or plant form,
Mr. Pierron brings to each object his fine sense of contemporary simplification and reorganization.

He has exploited the sharp crispness of form which is
natural to thin sheets of metal. His saw blade becomes the
drawing tool and all of the controlled precision of a fine
craftsman is reflected in each stroke. Using oxidation to
emphasize form, he dramatizes both the positive and the
negative shape to compose the total design.

1. "Jonah and the Whale," sterling silver pin; Robert
   Pierron.

2. Sterling silver pin; Robert Pierron.

3. "Rose," sterling silver pin; Robert Pierron.

4. "Owl," sterling silver pin; Robert Pierron.

5. "Jonah and the Whale," sterling silver pendant; Robert
   Pierron.

6. "Birds," sterling silver pin; Robert Pierron.

7. "Fish," sterling silver wire pendant; Robert Pierron.

2

3

4

6

5

7

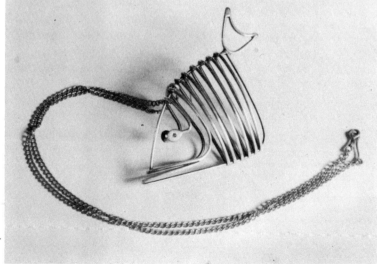

Scytho-Sarmatian bridle fittings

Fig. 212

into the frame of an object. Figure 212 shows how even the simple planes of a reindeer had to be enriched by translating them into parts of other animal forms. It should be noticed that not only were the objects carefully organized but also that the space between—the negatives formed by positive form—were carefully planned to create visual variety.

A DEFINITION: ABSTRACTION

Though terms vary, it might be said that there are two fundamental approaches to design: the *abstract* and the *nonobjective*.

The abstract might be defined as the invention of form based on a specific object. To abstract is to alter and reorganize reality. That there is a great range of possibilities within the term abstraction can be recognized when one considers that merely translating objects in the round to a two-dimensional surface, as in painting, is already an abstraction even if the rendition is photographically accurate. The artist has already made important decisions regarding what is to be used and what is to be discarded.

However, within the framework of abstraction one might take considerable liberties with form as it originally exists. An object may be simplified, dissected, and recombined. It may be enriched in shape and surface where, in its original form, these aspects might lack interest.

The principle of showing simultaneous views—the top, bottom, and sides or the inside and the outside of an object—has often been used in abstraction.

Emphasis of existing basic qualities such as the sharpness of projections or the softness of rounded forms may be important in stating the *essentials* of a natural form. Throughout man's creative history, these intellectual or intuitive processes have served to identify the artist with his time and with his environment.

The following illustrations can better demonstrate the above design approaches. It should be interesting to note that they vary considerably in time and geography. Surely this indicates that the natural development of abstract imagination is both individual and universal. (See Fig. 213.)

It is difficult for one who has not long thought in terms of abstraction to organize form, shape, and texture in a personal way. On the one hand it is all too easy to be influenced by the apparently successful solutions of other artists—in which case work becomes unnecessarily derivative. The alternative is too often an attempt at an obvious originality: an originality based on extremes of distortion and shock effect. If these extremes are not based on a deep knowledge of the original subject, they are most often trite in their efforts at novelty.

There is no easy formula for a sound abstract sense. It can only be the natural result of an all-consuming curiosity

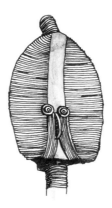

Ornamental shield,
Kerema District, New Guinea

Ritual object,
Gaboon, Africa

Pablo Picasso,
20th century painter

Fig. 213

Mayan war god,
Yucatán, Mexico

Eskimo dance mask, Alaska

Head of an Apostle,
early 13th century France

Fig. 213

Balinese shadow puppet

Zuñi dance mask,
S.W. United States

Paul Klee, 20th century painter

Fig. 213

191

about the nature of all things. The more you know about an object, the greater the latitude for intelligent interpretation and reorganization. At any point in the abstraction of form there should be visible the essential truth of the original form. It should be impossible to forget the bone, the muscle, and the sinew that lie beneath the skin. If, in reorganizing the concept of a snail shell, one totally ignores the pure geometry of its convolutions, the essence of the idea of a snail shell is lost.

Fortunately, we live today in an atmosphere of free inquiry. To enable us to delve into the complete knowledge an artist should seek, we have infinite sources of information. The artist must make use of the extensions of the naked eye; the microscope, the camera, the telescope are all extensions which give greater scope to our reality. They allow us to see into, through, and around things which are too small or too complex for the eye alone. Everything that these devices can bring to us is logical to use in creative expression.

The history of man's efforts in constructing meaningful images—his art—is a part of our total knowledge. It can and also should be used.

This does not mean that it is valid to repeat the forms of the past, only that it is possible and certainly right to analyze past approaches to abstraction. In the process of analysis, some of the historic solutions become the equipment of the artist in a natural way and can be used when the need occurs. Not the actualities of defined form but the ideas of abstraction may thus be used.

Turning one's back on the past—just because it is the past—discards unthinkingly the logic of development which has caused man to grow with each generation.

Perhaps the best equipment for an artist is a sketchbook into which he deposits daily that which intrigues him and stimulates his desire for greater knowledge. The mere act of defining an image or an idea on paper serves to fix these elements permanently in the mind.

A DEFINITION: THE NONOBJECTIVE

The nonobjective direction in design is, if anything, even more purely intellectual. By definition, nonobjective forms are *invented* to suit a need. They are the pure expression of emotion or intellect without the obvious starting point of an image (natural form).

An imagination based on the knowledge necessary to sound abstraction, as defined previously, is all-important to the invention of nonobjective form.

Because the possibilities of invention in this sense are so limitless, there is always the possibility of choosing the easy solution. The rules of the game are self imposed and have no social framework. The entire direction is so personal that many artists—who may lack depth in creative personality—cannot work in this direction comfortably. This is probably why so much nonobjective design is anonymous in its similarity.

As an approach to nonobjective invention, a few aspects may be safely defined.

The accidental form: often the accident of natural or man-made form may suggest the development of personal expression.

The texture of wood grain, the fracture planes of stone, or the silhouette of a hole in a windowpane can suggest a direction of interpretation and development.

An aerial photograph of a river delta or of farmland might suggest interplays of sinuous or angular shapes to be organized into satisfying images.

This is not to suggest that these photographic images should be translated directly into pictorial or sculptural forms. Rather, the perceptive imagination

1

2

3

Alice Boatright, relatively new to the field of jewelry making, is a fine example of the conscientious, imaginative craftsman working today. Though her formal training in techniques is limited, her enthusiasm and fine craftsmanship combine to create jewelry of individuality which would be a credit to earlier, more formal systems.

Miss Boatright has combined materials in new and exciting ways. In example 1 she has made an accurate impression of a mollusk shell in wax and has imbedded an irregular gem. After casting in silver, the result is of great textural interest. The cuff links in example 4 combine end grain bamboo and ivory, capturing the subtlety of one and the ornateness of the other. In example 5 she has paved an area with polished but irregular fragments of garnet to form earrings of great luster and originality.

Such inventive use of materials, forms, and textures makes contemporary jewelry a field of constant surprise and innovation.

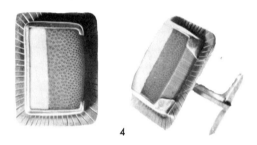

4

1. Pin, sterling silver and tumbled stone; Alice Boatright.

2. Pendant of silver, gold, and ebony; Alice Boatright.

3. Pin of silver, gold, and cat's eye shell; Alice Boatright.

4. Cuff links of ivory, bamboo, and silver; Alice Boatright.

5. Earrings of tumbled garnets and silver; Alice Boatright.

6. and 7. Ring of gold and tumbled stone, front and side views; Alice Boatright.

Photos by Alice Boatright

5

6   7

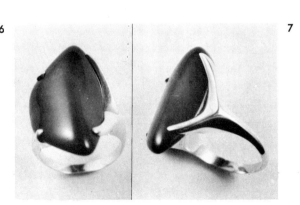

1

The author, perhaps because of early interests in drawing and natural history, has found that animal and human subjects, often reflecting delight in the abstractions of earlier times, are the most stimulating of many design directions. He has long been interested in using a variety of materials and both old and new techniques in the development of his jewelry, feeling that the greater the repertoire of experiences, the more flexible the approach to a design solution can be.

As a personal preference, the author creates only one example of each design though often many variations of a basic idea might be developed.

He has a strong conviction that a unique work in precious metal may incorporate all those elements which constitute a serious work of art in painting and sculpture, and, since this is the case, a sound training in a variety of media is of great importance to the contemporary artist-designer in jewelry.

1. "Fighting Cock," silver and gold pin, and "Wild Bull," silver and gold pin; Robert von Neumann.　　Photo by Hans van Nes

2. "Lions," cuff links in cast silver; Robert von Neumann.

3. "Hedgehog," silver pin, and "Reindeer," silver pin; Robert von Neumann.

4. "Roman Face," repoussé, chased and engraved silver pendant, and "The Pompeiian," fused silver pin; Robert von Neumann.

5. "Fish," silver pendant; Robert von Neumann.

6. "Jungle Fowl," silver and laminated metals pin; Robert von Neumann.

7. "King David," silver pin; Robert von Neumann.　　Photo by Hans van Nes

8. "Flounder," silver pin with ebony, ivory, and gold inlay; Robert von Neumann.

9. "Columbus," pendant in gold and silver; Robert von Neumann.

10. "Daphne," silver pendant. "Silver Perfume Bottle and Pendant." "King on a Throne," necklace and pendant of silver, gold, cloisonné enamel, and garnet; Robert von Neumann.　　Photo by Charles Uht

3

4

5

6

7

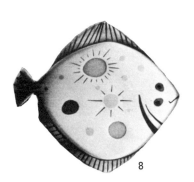

8

9

10

11

12

13

11. "The Gladiator," silver, gold, and laminated metals pin; Robert von Neumann.

12. "The Knight," silver pin with gold, turquoise, niello, and laminated metals; Robert von Neumann.

13. "Lancer," pendant in silver and gold; Robert von Neumann.

14. "Virtue," cast silver pendant; Robert von Neumann.

15. Bird pin, gold with blue and green cloisonné enamel and garnet eye; Robert von Neumann.

16. "Amazon," cast silver pendant with jasper shield; Robert von Neumann.

15

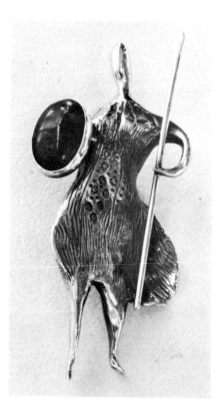

14

16

should be stimulated and fortified by constantly recognizing the usable suggestions they offer. Again it is a matter of seeing with greater clarity and receptivity.

Once the mind is organized into these terms, the emotional-intellectual evolution of nonobjective forms can become both personal and developmental. Equipped in this way, the artist can approach each problem as an exciting exercise in interpretation. Without such organization the result is sure to be eclectic, trite, and repetitive.

Should a form or organization of forms transmit the essence of sharpness, the shapes used would be the ultimate in sharpness as the artist feels sharpness to be.

He may decide that the over-all quality will be that of tension and disorganization and he will employ shape, color, and texture to best express this. There is a variety of these essential qualities that could be combined or exploited separately. Such feelings as softness, ornateness, simplicity, stability, fluidity, rigidity, could all be the framework over which the expression is built.

All of the basic principles of design—variety, repetition, tension, and relaxation—are needed to make a nonobjective form a thing of constant interest. (See Fig. 214.)

Each of these form inventions possesses a basic similarity—they are all compounded of curved lines terminating in points and also have one or more in-

Fig. 214

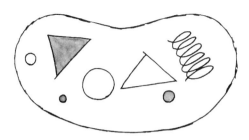

Fig. 215

ternal shapes. Some are obviously more pleasing to the eye than others. What causes pleasure in one and a sense of boredom or irritation in another?

In some the sheer variety of all parts creates an interest that is total. The positive—the form itself—is no more compelling than the negative space(s) formed by the limits or enclosure of the form. Fig. 214F is boring because there is not enough variety in proportion to make the surrounding form interesting. In this respect Fig. 214E is much more stimulating.

Figs. 214A, K, and J, in their essential symmetry, are exhausted of their shape excitement much more rapidly than are Figs. 214B, D, and L.

All of the elements within a confined form must relate intelligently to each other. A painful mistake, often seen in beginning shape inventions, is the uncomfortable combination of disparate shapes. In Fig. 215 the severe geometry of the haphazard internal shapes has nothing in common with the totally different—and uninspired—surrounding shape. Far from creating an interesting

tension, the result is only an unpleasant and unthinking chaos.

Tensions—the strong contrast of qualities within a single expression—can be very important. The following inventions (see Fig. 216) demonstrate a number of ways in which such tensions may be intellectually exploited.

Balance—already referred to—is of utmost importance. If a basically stable, and calm, form has a note of excitement somewhere within the form, this must be placed in such a position that the eye finds it in balance with the total shape. This may be achieved by repetition as well as single placement. (See Fig. 217.)

Another sort of balance is involved with touch or tangency. Since the discipline of containment is so necessary in art—what some artists call *structure*—anything that threatens containment should be intentional—never an accident.

In Fig. 218A, the single point tangent to the containing form destroys balance and creates an uncomfortable sense of leakage. In Fig. 218B, where several points touch the exterior, the central form interacts with the exterior in a sort of support—really balance.

Fig. 218C has only one form with a common edge. The eye is drawn to this point irresistibly, but without good reason. Fig. 218D has a better balance of shapes with a common edge.

The elements of balance and variety are equally important to abstract devel-

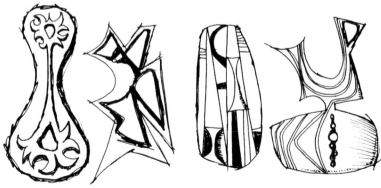

Fig. 216

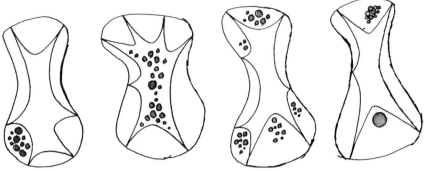

Fig. 217

opment and to nonobjective invention. Figs. 219, 220, and 221 indicate how variety may develop as an idea is reinterpreted many times.

The same search for the essential is necessary in abstraction, and without this an abstract form tends to become overly specific. An aspect of cuteness—with all of its transient shallowness—enters into abstraction when the superficial surface becomes more important than the essential identity of the object. As an example, much inexpensive costume jewelry is made in the shape of dogs, cats, clowns, and so forth. Invariably the designer has attempted to be too specific: it had to be an Irish setter, or a Siamese cat, or a Ringling clown. In being so specific the designer felt compelled to create an illusion of reality unsuited to the material even if well enough suited to a mass-production technique. The treatment of these forms is never subtle or exacting enough to bring it off as virtuosity. The result can only be a clumsy apeing of nature.

Is it not more fitting to have the material in mind as well as the technique of working the material, and then adapt the form-idea to these truths? Only in this way can an abstraction have meaning, since it must always be a restatement of fact expressed in a language imposed by the tools and the materials used.

Every point made so far could apply to any visual art process. Jewelry has, however, a special function which has its effect on design in this medium. Jewelry is decoration. Its only function is to enrich the appearance of the wearer, but pure decoration has sometimes been frowned upon in recent years. Perhaps this has been due to a great lack of per-

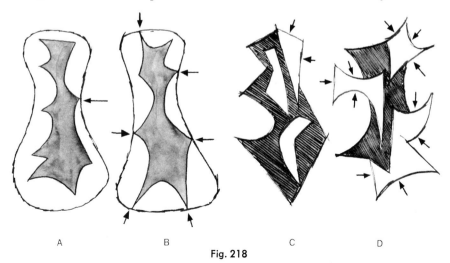

A   B   C   D

Fig. 218

Fig. 219

sonality in decorative art in the last century or so. Since decorative art has always had greater currency with the public than have the traditionally more personal forms of art, the decorative artist has, too often, adjusted his ideas to what he thought the public would want. This has led to the very conventional repetition of shape and treatment we still suffer today.

To be decorative, an object need not be ornate or complex. It needs only to capture the attention, then to stimulate curiosity and allow for leisurely contemplation of its uniqueness. It must, above all, be stimulating enough to bring attention to itself again and again. In this respect, all successful art is decorative—it enriches its environment as well as the visual experience and knowledge of the viewer.

The concept of individual statement is all-important in art, and especially so in jewelry. The successful piece of jewelry separates the wearer from the throng. If everyone wore the same design jewelry would soon be discarded.

In jewelry, as in all other art forms, craftsmanship can only implement and support sound design, never supplant it. Craftsmanship alone may well express virtuosity, but never invention and exploration.

Sound design without the support of

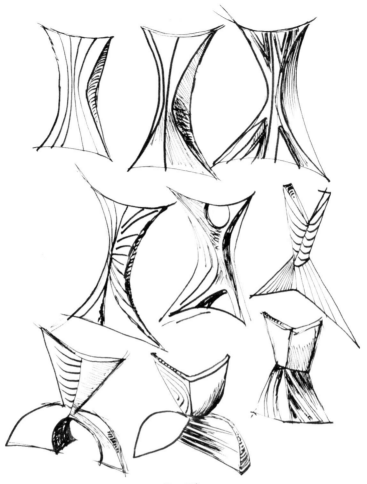

Fig. 220

excellent craftsmanship is equally inadequate, since slipshod appearance forcibly detracts from the total impact of a work of art. A mutual interdependence is not only necessary, it is the reason for all serious art. If possible, this is even more true of jewelry than of other forms. By its nature a piece of jewelry is precious, not only in its intrinsic value as gold or silver or gems, but also because of its smallness and the delicacy of its design and workmanship. Being small, it invites close scrutiny as well as a desire to touch and savor its form and surface. Such close scrutiny makes unthinking and unloving treatments of materials all too evident when they exist.

Many of the barriers to sound design have already been mentioned. One point requires additional emphasis. This may be defined as *design honesty*. Perhaps the ultimate sin in any art is to plagiarize ideas without adding something to them. The sin is first against the originator, whether a contemporary or one who lived and worked centuries ago. More important, the sin is against the plagiarist himself. He has diminished his own stature by depending completely on another's intellect. He has wasted to some degree his unique ability of decision and selectivity. In using the easy solution he has forfeited his right to pride, satisfaction, and all sense of achievement. This is the tragedy inherent in all how-to-do-it kits and formulated design manuals. Though the foregoing is true, it is also true that it is

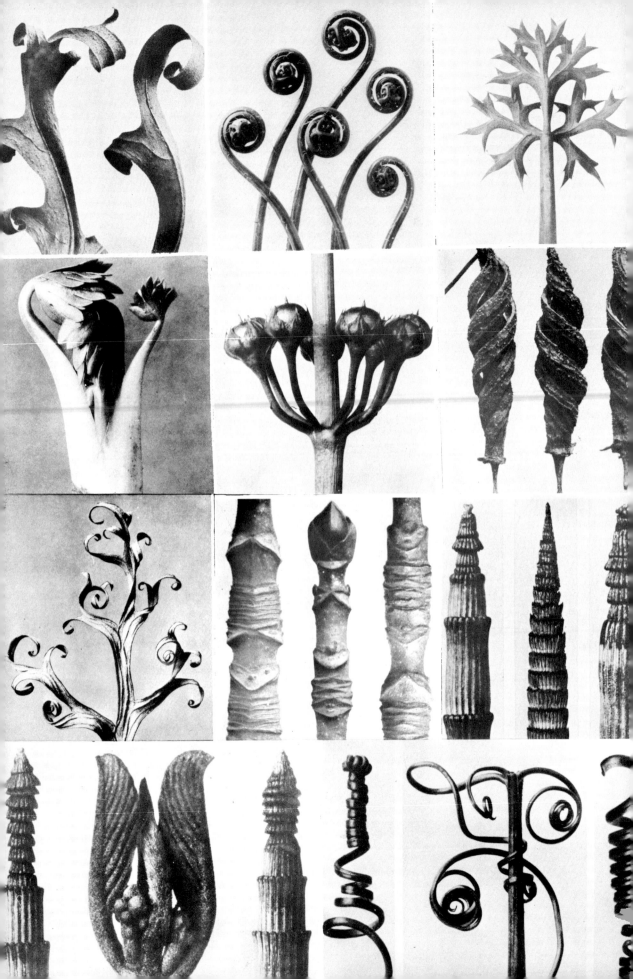

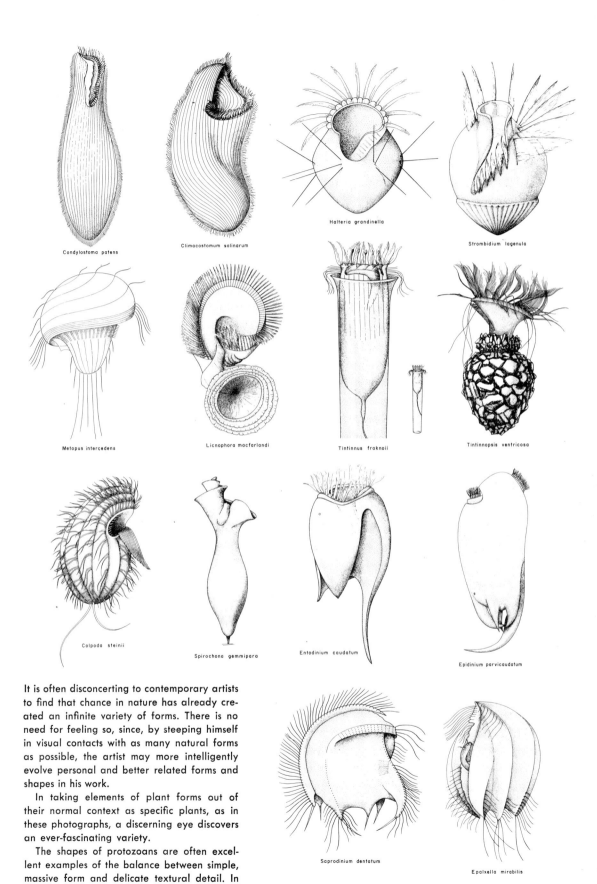

Condylostoma patens

Climacostomum salinarum

Halteria grandinella

Strombidium lagenula

Metopus intercedens

Licnophora macfarlandi

Tintinnus fraknoii

Tintinnopsis ventricosa

Colpoda steinii

Spirochona gemmipara

Entodinium caudatum

Epidinium parvicaudatum

Saprodinium dentatum

Epalxella mirabilis

It is often disconcerting to contemporary artists to find that chance in nature has already created an infinite variety of forms. There is no need for feeling so, since, by steeping himself in visual contacts with as many natural forms as possible, the artist may more intelligently evolve personal and better related forms and shapes in his work.

In taking elements of plant forms out of their normal context as specific plants, as in these photographs, a discerning eye discovers an ever-fascinating variety.

The shapes of protozoans are often excellent examples of the balance between simple, massive form and delicate textural detail. In seeing how living forms have solved this design problem, an artist may find clues to aid him in his own invention.

Drawings by Alice Boatright

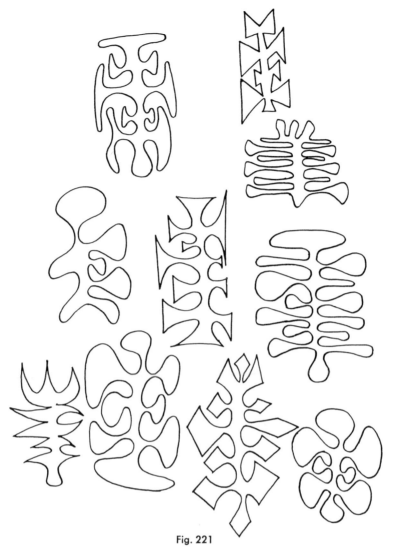

Fig. 221

impossible to invent form without bringing to the invention some of one's remembered visual experiences

What you have seen you own, and what you own you can use, but to use something exactly as you have seen it—without adding, or even subtracting, from it—deprives you of such ownership just as a museum copyist of an Old Master cannot lend his name to a faithful copy without identifying it as a copy.

There are, however, times when memory moves the hand without conscious acknowledgment of plan or reason. The echo of a former delight makes its sound quickly and quietly as a pencil might move from one point to another in a sketchbook.

It is important to make of memory a storehouse holding a multitude of shapes, surfaces, colors, and lines, as well as complete images. If this memory resource is full enough, it will become difficult to depend too greatly on another's ideas. Alternatives replace limitations; confidence replaces the contentment with being merely adequate.

Within the framework of this book it is possible to give only a hint of the many stimuli that could help to develop a design personality. The preceding photographs and drawings indicate the range of resources available, but anyone with a strong sense of curiosity will find it a simple matter to add to the list in variety and depth.

BIBLIOGRAPHY

JEWELRY TECHNIQUES (ENGLISH LANGUAGE)

Abbey, Staton: "The Goldsmith's and Silversmith's Handbook," London, Technical Press, Ltd., 1952.

Auld, J. Leslie: "Your Jewellery," London, Sylvan Press, 1951; Peoria, Ill., Chas. A. Bennett Co. (distributors).

Baxter, William T.: "Jewelry, Gem Cutting and Metalcraft," ed. 3, New York, McGraw-Hill Book Co., Inc., 1950.

Bovin, Murray: "Jewelry Making for Schools, Tradesmen and Craftsmen," New York, Murray Bovin, 68-36 108th St., Forest Hills, Long Island, 1952. Revised and enlarged, 1959.

Clegg, Helen, and Larom, Mary: "Jewelry Making for Fun and Profit," New York, David McKay Co., Inc., 1951.

Cuzner, Bernard: "Silversmith's Manual," London, N.A.G. Press, Ltd., 1935, 1949.

Dragunas, Andrew: "Creating Jewelry for Fun and Profit," New York, Harper and Brothers, 1947.

Handy and Harman: "Karat Golds for the Craftsman," New York, Handy and Harman, 82 Fulton St., 1951.

Kronquist, Emil F.: "Metalcraft and Jewelry," Peoria, Manual Arts Press, 1926.

Linick, Leslie L.: "Jeweler's Workshop Practices," Chicago, Henry Paulson and Co., 131 S. Wabash Ave., 1948.

Martin, Charles J., and D'Amico, Victor: "How to Make Modern Jewelry," Scranton, International Textbook Co., 1949.

Maryon, Herbert: "Metalwork and Enamelling," New York, Dover Publications, 1955.

Pack, Greta: "Chains and Beads," New York, D. Van Nostrand Co., Inc., 1952.

Pack, Greta: "Jewelry and Enameling," ed. 2, New York, D. Van Nostrand Co., Inc., 1953.

Rose, Augustus F., and Cirino, Antonio: "Jewelry Making and Design," Worcester, Mass., Davis Publications, Inc., revised, 1946.

Sanger, Arthur and Lucille: "Cabochon Jewelry Making," Peoria, Chas. A. Bennett Co., Inc., 1951.

Smith, F. R.: "Small Jewellery," London and New York, Pitman Publishing Corp., 1951.

Wiener, Louis: "Hand Made Jewelry," New York, D. Van Nostrand Co., Inc., 1948.

Wilson, Henry: "Silver Work and Jewellery," New York and London, Pitman Publishing Corp., 1902.

Winebrenner, D. Kenneth: "Jewelry Making as an Art Expression," Scranton, International Textbook Co., 1953.

JEWELRY TECHNIQUES (FOREIGN LANGUAGE)

Boitet, Alfred: "Traite Pratique du Bijoutier-Joaillier" (*French*), Paris, Editions Garnier Frères, 6 rue des Saints-Pères.

Braun-feldweg, Dr. Wilhelm: "Metall-Werkformen und Arbeitsweisen" (*German*), Verlag Gold und Silber, New York, Stechert-Hafner, Inc.

Czerwinski, Albert, and Hub, Friedrich: "Die Goldschmiedelehre" (*German*), Leipzig, Verlag Wilhelm Diebener, 1931.

Diebener Wilhelm: "Handbuch des Goldschmieds" (*German*), Leipzig, Verlag Wilhelm Diebener, 1929.

Herman, Reinhold: "Elementare Gestaltungslehre für den Goldschmied" (German), Verlag Gold und Silber, New York, Stechert-Hafner, Inc.

Schwan, Christian: "Die Metalle, Ihre Legierungen und Lote," "Die Oberflachen behandlung der Metalle," "Rezept und Werkstattbuch für den Gold- und Silberschmied" (German), Verlag Gold und Silber, New York, Stechert-Hafner, Inc.

Wilm, H. J.: "Lebendige Goldschmiedekunst" (German), Verlag Gold und Silber, New York, Stechert-Hafner, Inc.

HISTORIC TECHNIQUE REFERENCES

Bergsoe, Paul (English translation by F. C. Reynolds): "The Gilding Process and the Metallurgy of Copper and Lead Among the Pre-Columbian Indians," "The Metallurgy and Technology of Gold and Platinum Among the Pre-Columbian Indians" (References to granulation and other fusing techniques), Copenhagen, Danmarks Naturvidenskabelige Samfund, I Kommission Hos G.E.C. Gad. Vimmelskaftet 32, 1937, 1938.

Cellini, Benvenuto (English translation by C. H. Ashbee): "Treatises on the Arts of Goldsmithing and Sculpture."

Maryon, Herbert: "Metalwork and Enamelling" (References to granulation, lamination, inlay, etc.), New York, Dover Publications, Inc., 1955.

Milliken, William M.: "The Art of the Goldsmith," reprinted from the Journal of Aesthetics and Art Criticism, Vol. VI, No. 4, June 1948.

Rosenberg, Marc: "Geschichte der Goldschmiedekunst" (German), Vols. I, II, III (References to niello and granulation). Out of Print—Library of Congress, Class-NK 7106, Book-R8.

Wilson, Henry: "Silverwork and Jewellery" (References to inlay, lamination, and Japanese techniques), London and New York, Pitman Publishing Corp., 1902.

Withered, Newton: "Medieval Craftsmanship and the Modern Amateur," London, Longmans, Green & Co., Ltd., 1923.

HISTORY OF JEWELRY AND RELATED FORMS

Adair, John: "Navajo and Pueblo Silversmiths," Norman, University of Oklahoma Press, 1945

Alexander, Christine, Ed.: "Ancient Egyptian Jewelry," "Chinese Jewelry," "Greek and Etruscan Jewelry," "Jewelry, The Art of the Goldsmith in Classical Times," "Medieval Jewelry," "Near Eastern Jewelry," "Renaissance Jewelry," New York, Metropolitan Museum of Art.

Bainbridge, Henry C.: "Peter Paul Faberge, His Life's Work," London and New York, B. T. Batsford, Ltd., 1949.

Banco de la República, Bogotá: "80 Masterpieces from the Gold Museum," Colombia, Banco de la República, Bogotá, 1954.

Barradas, Jose Perez de: "Orfebreria Prehispánica de Columbia" (Spanish), (Excellent examples of Pre-Columbian goldwork), Colombia, Banco de la Republica, Bogotá, 1958.

Bradford, Ernle Dusgate Selby: "Contemporary Jewellery and Silver Design" (Examples of modern English commercial design), London, Heywood & Co., Ltd., 1950.

Bradford, Ernle Dusgate Selby: "Four Centuries of European Jewellery," New York, Philosophical Library, 1953.

Burch-Korrodi, Meinard: "Orfevrerie D'Eglise" (French), (Contemporary French examples of enameled ecclesiastic objects and jewelry), Paris, Editions Alsatia, 1956.

Burger, Dr. Willy: "Abendlaudische Schmelzarbeiten" (German), Berlin, Richard Carl Schmidt & Co., 1930.

Burgess, Frederick W.: "Antique Jewelry and Stones," New York, Tudor Publishing Co.

Carli, Enzo: "Pre-Conquest Goldsmith's Work of Colombia," New York, W. S. Heinman, 1958.

Evans, Joan: "A History of Jewelry, 1100–1870," New York, Pitman Publishing Corp., 1953.

Gehring, Prof. Oscar: "Josef Wilm, der Gold- und Silberschmied" (German), (Excellent contemporary work in granulation, etc.), Verlag Gold und Silber, New York, Stechert-Hafner, Inc.

Hald, Arthur: "Contemporary Swedish Design," Stockholm, Nordisk Rotogravyr, 1951.

Hara, Shinkichi: "Die Meister der Japanischen Schwertzieraten" (German), (Examples of Japanese inlay and patination), Hamburg, Verlag des Museums für Kunst und Gewerbe, 1931.

Hendley, Thomas H.: "Monograph on Indian Jewellery" (Jewelery of India and Ceylon), London, The Journal of Indian Art, Vol. 12, W. Griggs and Sons, 1884–1900.

Jessup, Ronald F.: "Anglo-Saxon Jewellery," New York, Frederick A. Praeger, Inc., 1953.

Jossic, Yvonne Françoise: "1050 Jewelry Designs" (*Some fine historic examples*), Philadelphia, Alfred Lampl, 1946.

Kelemen, Pal: "Medieval American Art, Vols. I & II" (*Excellent text and illustrations*), New York, The Macmillan Co., 1943.

McCarthy, James Remington: "Rings Through the Ages," New York, Harper & Brothers.

Muller-Erb, Rudolph: "Der Goldschmied Mohler" (*German*), Stuttgart, Chr. Belser Druckerei und Verlag, 1941.

Rogers, Frances, and Beard, Alice: "5000 Years of Gems and Jewelry," Philadelphia, J. B. Lippincott Co., 1947.

Rossi, Filippo: "Italian Jeweled Arts," New York, Harry N. Abrams, Inc., 1954.

Salin, Bernhard: "Die Altgermanische Thieronamentik" (*German*), Stockholm, Wahlstrom & Widstrand, Forlag.

Schweeger-Hefel, Annemarie: "Afrikanishe Bronzen" (*German*), (*West African design in metal from Ife and Ashanti areas*), Vienna, Kunstverlag Wolfrum, 1948.

Steingraber, E.: "Antique Jewelry" (*Excellent examples of medieval and Renaissance European jewelry*), New York, Frederick A. Praeger, Inc., 1957.

Thoma, Hans: "Kronen und Kleinodien" (*German*), Berlin, Deutscher Kunstverlag, 1955.

Ugglas, Carl Gustaf: "Kyrkligt Guld-och Silversmide" (*Swedish*), Stockholm, Wahlstrom & Widstrand, Forlag, 1933.

Woodward, Arthur: "A Brief History of Navajo Silversmithing," Flagstaff, Northern Arizona Society of Science and Art, 1938.

LAPIDARY INFORMATION

Anderson, B. W.: "Gem Testing for Jewelers," London, Heywood & Co., Ltd., 1947.

Drake, Dr. E. H., and Pearl, R. M.: "The Art of Gem Cutting," Portland, Mineralogist Publishing Co., 1945.

Gravender, Milton F.: "Fascinating Facts about Gems," Los Angeles, Gemological Institute of America.

Howard, J. Harry: "Revised Lapidary Handbook," Greenville, S. C., 504 Crescent St., 1946.

Howard, Henry: "The Working of Semi-Precious Stones," Greenville, S. C., 504 Crescent St.

Kraus, E. H., and Slawson, C. B.: "Gems and Gem Material," ed. 5, New York, McGraw-Hill Book Co., Inc., 1947.

Shipley, Robert M.: "Dictionary of Gems and Jewelry," Los Angeles, Gemological Institute of America.

Sinkankas, A.: "Gem Cutting, a Lapidary's Manual," New York, D. Van Nostrand Co., Inc., 1955.

Sperisen, Francis J.: "The Art of the Lapidary," Milwaukee, Bruce Publishing Co., 1950.

Willems, J. Daniel: "Gem Cutting," Peoria, Manual Arts Press, 1948.

ENAMELING TECHNIQUES AND HISTORY

Bates, Kenneth F.: "Enameling, Principles and Practice," Cleveland, World Book Co., 1951.

Cunynghame, H. H.: "Art Enamelling on Metals," London, Constable & Co., Ltd., 1906.

Dalton, O. M.: "Byzantine Enamels in Mr. Pierpont Morgan's Collection" (*Fine examples of early European cloisonné and champlevé techniques*), London, Chatto and Windus, Ltd., 1912.

Day, Lewis F.: "Enamelling," London and New York, B. T. Batsford, Ltd., 1907.

Gauthier, Marie-Madeleine S.: "Emaux Limousins Champlevés des XII, XIII et XVI Siècles" (*French*), Paris, Gérard Le Prat, 268 Boulevard Saint-Germain, 1950.

Hasenohr, Curt: "Email" (*German*), Dresden, Verlag der Kunst, 1955.

Koningh, H. de: "Preparation of Precious and Other Metal Work for Enamelling," London, The Technical Press, Ltd., 1947.

Lavendau, Pierre: "Leonard Limousin et Les Emailleurs Française" (*French*), Paris, Henri Laurens.

Millenet, Louis-Elie: "Enamelling on Metal," London, The Technical Press, Ltd., 1947.

Otten, Mitzi, and Berl, Kathe: "The Art of Enameling," New York, 1947 Broadway, 1950.

Thompson, Thomas E.: "Enameling on Copper and Other Metals," Highland Park, Ill., Thomas C. Thompson Co., 1539 Deerfield Road, 1950.

Untracht, Oppi: "Enameling on Metal," Philadelphia, Chilton Company—Book Division, 1957.

ENGRAVING TECHNIQUES

Bowman, John J.: "Jewelry Engraver's Manual," New York, D. Van Nostrand Co., Inc., 1954.

PLASTICS TECHNIQUES

Castolite Company: Descriptive process Bulletins published as "The Castoliter," Woodstock, Ill., The Castolite Company.

BOOKS OF INTEREST IN THE ABSTRACTION OF NATURAL FORMS

Bliss, Robert Woods Collection: "Indigenous Art of the Americas," Washington, Smithsonian Institution, 1947.

Boas, Franz: "Primitive Art," New York, Capitol Publishing Co., Inc., 1951.

Borovka, Gregory: "Scythian Art," New York, Frederick A. Stokes Co., 1928 (*out of print*).

Bossert, Helmuth T.: "Alt Syrien" (*German*), Tübingen, Ernst Wasmuth, 1951.

Bossert, Helmuth T.: "Art of Ancient Crete," London, A. Zwemmer, Ltd., 1937.

Einstein, Carl: "Afrikanische Plastik" (*German*), Berlin, Verlag Ernst Wasmuth A.G., 1921.

Elkin, A. P.: "Art in Arnhemland," Chicago, University of Chicago Press, 1950.

Griaule, Marcel: "Folk Art of Africa," New York, Tudor Publishing Co., 1950.

Heine-Geldern, Robert von: "Indonesian Art," New York, The Asia Institute, 1948.

Leenhardt, Maurice: "Folk Art of Oceania," New York, Tudor Publishing Co., 1950.

Linton, Ralph, and Wingert, Paul S.: "Arts of the South Seas," New York, Museum of Modern Art, Simon and Schuster, Inc., 1946.

Lothrop, Samuel Kirkland: "Cocle, An Archaeological Study of Central Panama," Cambridge, Peabody Museum, Harvard University, 1937.

Markman, Sidney David: "The Horse in Greek Art," Baltimore, Johns Hopkins Press, 1943.

Minns, Ellis Hovell: "The Art of the Northern Nomads," London, H. Milford, 1944.

Roes, Anna: "Greek Geometric Art," Haarlem, 1933.

Roth, Edward: "Primitive Art from Benin," *International Studio Magazine,* January 1899.

Schmalenback, Wilhelm: "Die Kunst Afrikas" (*German*), Basle, Holbein Verlag, 1953.

Tischner, Herbert: "Oceanic Art," New York, Pantheon Books, Inc., 1954.

PERIODICALS

Ceramics Monthly (Sections on enameling) Professional Publications, Inc., Columbus, Ohio.

Craft Horizons (General craft information) Craft Horizons, Inc., 601 Fifth Avenue, New York 17, New York.

Crafter's Friend (Sections on enameling) The Potter's Wheel, Inc., 11447 Euclid Ave., Cleveland 6, Ohio.

Design Quarterly (Two issues on contemporary American jewelry, 1955 and 1959) Walker Art Center, 1710 Lyndale Avenue South, Minneapolis 3, Minnesota.

Deutsche Goldschmiede-Zeitung (German) Rühle-Diebener Verlag, Stuttgart, Olgastrasse 110.

Gold und Silber (German) Stechert-Hafner, Inc., 31 East 10th St., New York (distributors).

The Lapidary Journal Del Mar, California.

SUPPLY SOURCES FOR TOOLS AND MATERIALS

PRECIOUS METALS

Eastern Smelting & Refining Corp.
  107 W. Brookline St., Boston 18, Mass.
Goldsmith Brothers Smelting & Refining Co.
  58 E. Washington St., Chicago 2, Ill.
T. B. Hagstoz & Son
  709 Sansom St., Philadelphia 6, Pa.
Handy & Harman
  Bridgeport 1, Conn.
  82 Fulton St., New York 38, N. Y.
  425 Richmond St., Providence 3, R. I.
  1900 W. Kinzie St., Chicago 22, Ill.
  330 N. Gibson St., El Monte, Calif.
  141 John St., Toronto, Canada.
Hauser & Miller
  4011 Forest Park Blvd., St. Louis 8, Mo.
Southwest Refining Co.
  P.O. Box 2010, Dallas 21, Tex.
Wildberg Bros. Smelting & Refining Co.
  635 S. Hill St., Los Angeles 14, Calif.
  742 Market St., San Francisco 2, Calif.

COPPER, BRASS, AND BRONZE

T. E. Conklin Brass & Copper Co., Inc.
  54 Lafayette St., New York 13, N. Y.
Patterson Bros.
  15 Park Row, New York 7, N. Y.
Revere Copper & Brass, Inc.
  230 Park Ave., New York 17, N. Y.

FINDINGS

Allcraft Tool & Supply Co., Inc.
  15 W. 45th St., New York 36, N. Y.
C. E. Marshall Co.
  Box 7737, Chicago, Ill.

The Newall Manufacturing Company
  139 N. Wabash Ave., Chicago 2, Ill.
Wm. J. Orkin, Inc.
  373 Washington St., Boston 8, Mass.
Wildberg Bros. Smelting & Refining Co.
  635 S. Hill St., Los Angeles 14, Calif.
  742 Market St., San Francisco 2, Calif.

TOOLS

Allcraft Tool & Supply Co., Inc.
  15 W. 45th St., New York 36, N.Y.
Anchor Tool & Supply Co., Inc.
  12 John St., New York 7, N. Y.
William Dixon, Inc.
  32 E. Kinney St., Newark 1, N. J.
Paul H. Gesswein & Co., Inc.
  35 Maiden Lane, New York 7, N. Y.
Ernest Linick & Co.
  5 S. Wabash Ave., Chicago 3, Ill.

CHEMICAL SUPPLIES

The following list is a partial geographic distribution of chemical supply firms which are able to supply small amounts of chemicals to individuals.

East

Burrell Corp.
  2223 Fifth Avenue, Pittsburgh 19, Pa.
Howe & French, Inc.
  99 Broad St., Boston 10, Mass.
New York Laboratory Supply Co.
  78 Varick St., New York 13, N. Y.
Seidler Chemical & Supply Co.
  12–16 Orange St., Newark, N. J.

South

W. H. Curtin & Co.
  P.O. Box 606, Jacksonville, Fla.
  P.O. Box 1491, New Orleans 13, La.
Will Corporation of Georgia
  P.O. Box 966, Atlanta 1, Ga.

Midwest

Harshaw Scientific Co.
  1945 E. 97th St., Cleveland 6, Ohio.
  9240 Hubbell Ave., Detroit 28, Mich.
Kansas City Laboratory Supply Co.
  307 Westport Rd., Kansas City 11, Kans.
Physician's Hospital Supply Co.
  1400 Harmon Pl., Minneapolis 3, Minn.
Roemer-Karrer Co.
  810 N. Plankinton Ave., Milwaukee 3, Wis.
E. H. Sargent & Co.
  4647 W. Foster Ave., Chicago 30, Ill.

West

Braun-Knecht-Heimann Co.
  2301 Blake St., Denver 17, Colo.
  1400 16th St., San Francisco 19, Calif.
W. H. Curtin & Co.
  1812 Griffin St., Dallas 2, Tex.
Scientific Supplies Co.
  600 Spokane St., Seattle 4, Wash.

Canada

Canadian Laboratory Supplies, Ltd.
  403 St. Paul St., W., Montreal 1, Quebec.
  3701 Dundas St., W., Toronto 9, Ontario.
Cane & Co., Ltd.
  1050 W. 6th Ave., Vancouver, B. C.

CASTING EQUIPMENT AND SUPPLIES

The Cleveland Dental Mfg. Co.
  Cleveland, Ohio.
The Jelrus Company, Inc.
  136 W. 52nd St., New York 19, N. Y.
Kerr Dental Mfg. Co. (Wholesale only. Will
  supply addresses of local distributors)
  6081–6095 Twelfth St., Detroit 8, Mich.
Alexander Saunders & Co.
  95 Bedford St., New York 14, N. Y.
S. S. White Dental Mfg. Co.
  55 E. Washington St., Chicago 2, Ill.

LAPIDARY EQUIPMENT

M.D.R. Manufacturing Co.
  4853 W. Jefferson Blvd., Los Angeles 16, Calif.
Technicraft Lapidaries Corp.
  3560 Broadway, New York 31, N. Y.
Vreeland Manufacturing Co.
  4105 N. E. 68th Ave., Portland 13, Ore.

ELECTROPLATING EQUIPMENT

Hoover & Strong, Inc.
  111 W. Tupper St., Buffalo 1, N. Y.

GEMS

Polished

John J. Barry Co.
  P.O. Box 15, Detroit 31, Mich.
International Gem Co.
  15 Maiden Lane, New York 7, N. Y.
Sam Kramer
  29 W. 8th St., New York 11, N. Y.
William Mercer
  665 Fifth Ave., New York 22, N. Y.

Roughs

Grieger's
  1633 E. Walnut St., Pasadena 4, Calif.
V. D. Hill
  Route 7, Box 188, Salem, Ore.
Technicraft Lapidaries Corp.
  3560 Broadway, New York 31, N. Y.

ENAMELS

B. F. Drakenfeld & Co.
  45 Park Place, New York 7, N. Y.
Thomas C. Thompson
  1539 Deerfield Rd., Highland Park, Ill.

WOOD

Craftsman Wood Service Co.
  2727 S. Mary St., Chicago 8, Ill.
Sam Kramer
  29 W. 8th St., New York 11, N. Y.

PLASTICS

Cadillac Plastic & Chemical Co.
  727 W. Lake St., Chicago 6, Ill.
Castolite Co.
  Woodstock, Ill.
Rohm & Haas Company
  222 W. Washington Sq., Philadelphia 5, Pa.

IVORY AND OTHER UNUSUAL MATERIALS

Sam Kramer
  29 W. 8th St., New York 11, N. Y.

JEWELRY BOXES

Pictorial Box Co.
  Aurora, Ill.

APPENDIX

Troy Weight—used in weighing the precious metals.

   24 grains = 1 pennyweight (dwt.)
   20 dwt. = 1 ounce troy
   12 ounces = 1 pound troy
 5,760 grains – 1 pound troy

Avoirdupois Weight—used in weighing base metals.

  16 drams (or drachms) = 1 ounce Avoir.
  16 ounces = 1 pound Avoir.
  16 ounces = 7,000 grains
  28 pounds = 1 quarter
   4 quarters = 1 hundredweight (cwt.)
  20 hundredweight = 1 ton Avoir.

To convert ounces troy to ounces avoirdupois, multiply by 1.09714.
To convert ounces avoirdupois to ounces troy, multiply by 0.91146.

Gram Weight—basic unit of weight in the metric system.

   1 gram – 15.43 grains troy
   1.555 grams = 1 pennyweight (dwt.)
  31.104 grams = 1 ounce troy
  28.35 grams = 1 ounce Avoir.

Carat Weight—used in weighing precious and semiprecious stones. (The term *Karat* refers to the quality of purity in gold.)

  1 carat = $3\frac{1}{16}$ grains troy
  1 carat = .007 ounce Avoir.
  1 carat = $\frac{1}{5}$ gram

The carat is further divided into *points* for simple measurement:

   1 carat = 100 points
  $\frac{1}{2}$ carat = $5\%_{100}$ points
  $\frac{1}{4}$ carat = 25_{100} points
  $\frac{1}{8}$ carat = $12\frac{1}{2}/100$ points

COMPARATIVE WEIGHTS

1 ounce Avoir. = 0.912 ounce troy = 28.35 grams

16 ounces Avoir. = 14.6 ounces troy = 1 pound Avoir.

1 ounce troy = 480 grains = 31.1 grams

1 ounce troy = 20 pennyweight = 1.1 ounces Avoir.

1 pennyweight (dwt.) = 24 grains = 1.555 grams

1 pound Avoir. = 453.6 grams = 7,000 grains = 16 ounces Avoir.

1 pound troy = 373.2 grams = 5,760 grains = 12 ounces troy

1 dram = 60 grains = 3.888 grams

1 gram = 15.43 grains = 0.032 ounce troy

1,000 grams (1 Kilogram) = 2.2 pounds = 35.26 ounces Avoir.

1 grain = 0.065 gram

100 grains = 6.5 grams

1 carat = 1.5 grams = 3.086 grains = 0.007 ounce Avoir.

Apothecaries' Weight—The grain, the ounce, and the pound of the apothecaries' weight system are the same as for the troy weight system.

FLUID MEASURES

1 ounce = 29.57 cubic centimeters = 1.8 cubic inches

1 dram = $\frac{1}{8}$ ounce (0.125 ounce) (fluid)

1 gill = 4 ounces (fluid)

1 pint = 16 ounces (fluid)

1 quart = 2 pints = $\frac{1}{4}$ gallon = $57\frac{3}{4}$ cubic inches

1 gallon = 4 quarts = 128 ounces (fluid) = 3.78 liters and 231 cubic inches = 0.134 cubic foot

1 cubic centimeter (cc.) = 16.23 minims

1 liter = 1,000 cc. (a little more than 1 quart U.S.) = 0.264 U.S. gallon

1 cubic foot = 7.481 U.S. gallons = 1,728 cubic inches

1 Imperial gallon = 1.2 U.S. gallons = 4.54 liters = 277.27 cubic inches

MEASURES OF LENGTH

1 inch (1″) = 2.54 centimeters = 25.4 millimeters

1 foot (1′) = 0.305 meter = 30.48 centimeters = 304.8 millimeters

1 meter = 39.37 inches

1 centimeter = 10 millimeters

10 centimeters = 1 decimeter = 100 millimeters

100 centimeters = 1 meter = 10 decimeters

CIRCUMFERENCE

Diameter (Inches)	Circumference (Inches)	Diameter (Inches)	Circumference (Inches)
3	$9\frac{3}{8}$	8	$25\frac{1}{8}$
$3\frac{1}{2}$	$10\frac{5}{16}$	$8\frac{1}{2}$	$26\frac{11}{16}$
4	$12\frac{9}{16}$	9	$28\frac{1}{4}$
$4\frac{1}{2}$	$14\frac{1}{8}$	$9\frac{1}{2}$	$29\frac{13}{16}$
5	$15\frac{11}{16}$	10	$31\frac{3}{8}$
$5\frac{1}{2}$	$17\frac{1}{4}$	$10\frac{1}{2}$	$32\frac{15}{16}$
6	$18\frac{13}{16}$	11	$34\frac{1}{2}$
$6\frac{1}{2}$	$20\frac{3}{8}$	$11\frac{1}{2}$	$36\frac{1}{8}$
7	$21\frac{15}{16}$	12	$37\frac{11}{16}$
$7\frac{1}{2}$	$23\frac{1}{2}$		

SURFACE SPEEDS OF WHEELS
In feet per minute [f.p.m.]

Motor or Spindle Speed r.p.m.	2″	4″	6″	8″	10″	12″	14″
1,000	525	1,050	1,575	2,100	2,600	3,100	3,600
1,200	630	1,260	1,950	2,550	3,200	3,750	4,400
1,400	730	1,470	2,250	2,950	3,650	4,400	5,100
1,600	840	1,680	2,550	3,400	4,200	5,000	5,900
1,800	940	1,890	2,900	3,800	4,750	5,650	6,600
2,000	1,050	2,100	3,200	4,200	5,250	6,250	7,300
2,200	1,150	2,300	3,450	4,550	5,750	6,900	8,000
2,400	1,260	2,500	3,750	5,000	6,300	7,500	8,800
2,600	1,360	2,700	4,100	5,450	6,800	8,200	9,600
2,800	1,470	2,950	4,400	5,900	7,400	8,900	10,400
3,000	1,570	3,140	4,700	6,250	7,900	9,400	11,200
3,200	1,680	3,350	5,000	6,650	8,400	10,000	11,900
3,400	1,780	3,560	5,250	7,000	8,900	10,600	12,600
3,600	1,880	3,780	5,600	7,500	9,500	11,300	13,300

TEMPERATURE CONVERSIONS

To convert degrees Fahrenheit (°F) to degrees centigrade (°C), first subtract 32, then take 5/9 of the remainder. To convert degrees centigrade to degrees Fahrenheit, first multiply by 9/5, then add 32.

or

To convert degrees centigrade to degrees Fahrenheit, first multiply by 1.8, then add 32. To convert degrees Fahrenheit to degrees centigrade, first subtract 32, then divide by 1.8.

or

To convert degrees centigrade to degrees Fahrenheit, first multiply the centigrade figure by 9, then divide the obtained figure by 5 and add 32. To convert degrees Fahrenheit to degrees centigrade, first subtract 32 and multiply by 5, then divide the obtained figure by 9.

Each 1° C = 1.8° F. The number 32 represents the difference between the nominal starting points 0 and 32.

A Few Comparisons

°C		°F
1000	=	1832
500	=	932
100	=	212
0	=	32

RING SIZES

Each ring size unit differs 0.032″ from the next full size in diameter. The diameter of a ring is measured at the inside diameter at the center of the band width.

Size		Inch	Size		Inch
0	=	0.458″ dia.	6½	=	0.666″ dia.
¼	=	.466	7	=	.682
½	=	.474	7½	=	.698
¾	=	.482	8	=	.714
1	=	.490	8½	=	.730
1½	=	.506	9	=	.746
2	=	.522	9½	=	.762
2½	=	.538	10	=	.778
3	=	.554	10½	=	.794
3½	=	.570	11	=	.810
4	=	.586	11½	=	.826
4½	=	.602	12	=	.842
5	=	.618	12½	=	.858
5½	=	.634	13	=	.874
6	=	.650	13½	=	.890

FRACTIONAL AND DECIMAL INCHES TO MILLIMETERS

Fractions	Decimal Inches	Millimeters
1/64	0.0156	0.3969
1/32	0.0313	0.7937
3/64	0.0469	1.1906
1/16	0.0625	1.5875
5/64	0.0781	1.9843
3/32	0.0937	2.3812
7/64	0.1094	2.7781
1/8	0.1250	3.1750
9/64	0.1406	3.5718
5/32	0.1562	3.9687
11/64	0.1719	4.3656
3/16	0.1875	4.7624
13/64	0.2031	5.1593
7/32	0.2187	5.5562
15/64	0.2344	5.9530
1/4	0.2500	6.3499
17/64	0.2656	6.7468
9/32	0.2812	7.1437
19/64	0.2969	7.5405
5/16	0.3125	7.9374
21/64	0.3281	8.3343
11/32	0.3438	8.7312
23/64	0.3594	9.1280
3/8	0.3750	9.5249
25/64	0.3906	9.9217
13/32	0.4062	10.3186
27/64	0.4219	10.7155
7/16	0.4375	11.1124
29/64	0.4531	11.5092
15/32	0.4687	11.9061
31/64	0.4844	12.3030
1/2	0.5000	12.6999

ROUND WIRE *

Weight in Pennyweights or Ounces Per Foot in B and S Gauge

B & S Gauge	Thickness in Inches	Fine Silver Ozs.	Sterling Silver Ozs.	Coin Silver Ozs.	Fine Gold Dwts.	10K Yel. Gold Dwts.	14K Yel. Gold Dwts.	18K Yel. Gold Dwts.	Platinum Ozs.	Palladium Ozs.
1	.28930	4.38	4.32	4.30	161.0	96.2	109.	130.	8.91	4.99
2	.25763	3.47	3.43	3.41	128.	76.3	86.1	104.	7.07	3.94
3	.22942	2.75	2.72	2.70	101.	60.5	68.3	81.5	5.61	3.19
4	.20431	2.18	2.15	2.14	80.3	48.0	54.2	64.6	4.45	2.42
5	.18194	1.73	1.71	1.70	63.6	38.0	43.0	51.2	3.53	1.97
6	.16202	1.37	1.36	1.35	50.5	30.2	34.1	40.6	2.80	1.56
7	.14428	1.09	1.07	1.07	40.0	23.9	27.0	32.2	2.22	1.24
8	.12849	.863	.852	.848	31.7	19.0	21.4	25.6	1.76	.984
9	.11443	.685	.676	.673	25.2	15.1	17.0	20.3	1.39	.780
10	.10189	.543	.536	.533	20.0	11.9	13.5	16.1	1.11	.619
11	.09074	.431	.425	.423	15.8	9.46	10.7	12.7	.877	.491
12	.08080	.341	.337	.335	12.6	7.50	8.47	10.1	.695	.389
13	.07196	.271	.267	.266	9.96	5.95	6.72	8.01	.552	.309
14	.06408	.215	.212	.211	7.89	4.72	5.33	6.36	.437	.495
15	.05706	.170	.168	.167	6.26	3.74	4.23	5.04	.347	.154

16	.05082	.135	.133	.133	4.97	2.97	3.35	4.00	.275	.154
17	.04525	.107	.106	.105	3.94	2.35	2.66	3.17	.218	.122
18	.04030	.0849	.0838	.0834	3.12	1.87	2.11	2.51	.173	.0968
19	.03589	.0674	.0665	.0662	2.48	1.48	1.67	1.99	.137	.0767
20	.03196	.0534	.0527	.0525	1.96	1.17	1.33	1.58	.109	.0609
21	.02846	.0424	.0418	.0416	1.56	.931	1.05	1.25	.0863	.0483
22	.02534	.0336	.0331	.0330	1.23	.738	.833	.994	.0684	.0383
23	.02257	.0266	.0263	.0262	.979	.585	.661	.789	.0543	.0304
24	.02010	.0211	.0209	.0208	.777	.464	.524	.625	.0430	.0241
25	.01790	.0168	.0165	.0165	.616	.368	.416	.496	.0341	.0191
26	.01594	.0133	.0131	.0131	.489	.292	.330	.393	.0271	.0151
27	.01419	.0105	.0104	.0103	.387	.231	.261	.312	.0214	.0120
28	.01264	.00835	.00825	.00821	.307	.184	.207	.247	.0170	.00952
29	.01125	.00662	.00653	.00650	.243	.145	.164	.196	.0135	.00754
30	.01002	.00525	.00518	.00516	.193	.115	.130	.155	.0107	.00598
31	.00892	.00416	.00411	.00409	.153	.0914	.103	.123	.00847	.00474
32	.00795	.00330	.00326	.00325	.122	.0726	.0820	.0978	.00673	.00377
33	.00708	.00262	.00259	.00258	.0964	.0576	.0651	.0776	.00534	.00299
34	.00630	.00208	.00205	.00204	.0763	.0456	.0515	.0614	.00423	.00236
35	.00561	.00165	.00162	.00162	.0605	.0362	.0408	.0487	.00335	.00188
36	.00500	.00131	.00129	.00128	.0481	.0287	.0324	.0387	.00266	.00149
37	.00445	.00104	.00102	.00102	.0381	.0228	.0257	.0306	.00211	.00118
38	.00396	.000820	.000809	.000806	.0302	.0180	.0204	.0243	.00167	.000934
39	.00353	.000652	.000643	.000640	.0240	.0143	.0162	.0193	.00133	.000742
40	.00314	.000516	.000509	.000507	.0190	.0113	.0128	.0153	.00105	.000587

* Square wire is 1.27324 times as heavy as round wire of the same gauge.

DECIMAL EQUIVALENTS OF DRILL SIZES

Size	Decimal Equivalent	Size	Decimal Equivalent	Size	Decimal Equivalent
½	0.500	3	0.213	3/32	0.0937
31/64	.4843	4	.209	42	.0935
15/32	.4687	5	.2055	43	.089
29/64	.4531	6	.204	44	.086
7/16	.4375	13/64	.2031	45	.082
27/64	.4218	7	.201	46	.081
Z	.413	8	.199	47	.0785
13/32	.4062	9	.196	5/64	.0781
Y	.404	10	.1935	48	.076
X	.397	11	.191	49	.073
25/64	.3906	12	.189	50	.070
W	.386	3/16	.1875	51	.067
V	.377	13	.185	52	.0635
3/8	.375	14	.182	1/16	.0625
U	.368	15	.180	53	.0595
23/64	.3593	16	.177	54	.055
T	.358	17	.173	55	.052
S	.348	11/64	.1718	3/64	.0468
11/32	.3437	18	.1695	56	.0465
R	.339	19	.166	57	.043
Q	.332	20	.161	58	.042
21/64	.3281	21	.159	59	.041
P	.323	22	.157	60	.040
O	.316	5/32	.1562	61	.039
5/16	.3125	23	.154	62	.038
N	.302	24	.152	63	.037
19/64	.2968	25	.1495	64	.036
M	.295	26	.147	65	.035
L	.290	27	.144	66	.043
9/32	.2812	9/64	.1406	1/32	.0312
K	.281	28	.1405	67	.032
J	.277	29	.136	68	.031
I	.272	30	.1285	69	.029
H	.266	1/8	.125	70	.028
17/64	.2656	31	.120	71	.026
G	.261	32	.116	72	.025
F	.257	33	.113	73	.024
E—1/4	.250	34	.111	74	.0225
D	.246	35	.110	75	.021
C	.242	7/64	.1093	76	.020
B	.238	36	.1065	77	.018
15/64	.2343	37	.104	1/64	.0156
A	.234	38	.1015	78	.016
1	.228	39	.0995	79	.0145
2	.221	40	.098	80	.0135
7/32	.2187	41	.096		

MOHS SCALE

The Mohs Scale is a system for classifying the hardness of minerals. The following list includes some of the better-known gem materials as well as common comparisons:

Graphite, Talc	1		
		1½	Human skin
Gypsum (Plaster), Alabaster	2		
		2½	Fingernail
Calcite, Limestone, Mexican Onyx, Pearl	3		
		3½	Copper coin
Fluorite	4		
		4½	Lead glass
Apatite	5		
		5½	Window glass
Feldspar	6		
		6¾	Tungsten, chromium, carbon steel
Quartz	7		
Tourmaline, Zircon	7½		
Topaz	8		
Chrysoberyl, Beryl	8½		
Corundum (Rubies, Sapphires)	9		
Diamond	10		

Diamonds vary in hardness with location. In order of hardness:
Australia-Borneo, Hardest   South America   India   Africa, Softest

MELTING POINTS AND SPECIFIC GRAVITY OF PRINCIPAL NONFERROUS METALS

Metal	Melting Point Fahrenheit	Melting Point Centigrade	Specific Gravity
Platinum	3224	1773	21.45
Nickel	2645	1452	8.85
Copper	1981	1083	8.93
Gold	1945	1063	19.36
Silver	1761	962	10.56
Sterling silver	1640	893	10.40
Zinc	787	419	7.14
Lead	621	327	11.37
Tin	450	232	7.29

COMMON AND CHEMICAL NAMES OF COMPOUNDS *

Common Name	Chemical Name
Acetic ether	Ethyl acetate
Acid of sugar	Oxalic acid
Aldehyde	Acetaldehyde
Alum	
Alum flour	Generally refers to potassium aluminum sulfate
Alum meal	
Alumina	Aluminum oxide
Alumino-ferric	A mixture of aluminum and sodium sulfates
Alundum	Fused alumina
Aniline	Phenyl amine
Aniline salt	Aniline hydrochloride
Antichlor	Sodium thiosulfate
Antifebrin	Acetanilide
Antimony black	Antimony trisulfide
Antimony bloom	Antimony trioxide
Antimony glance	Antimony trisulfide
Antimony red	
Antimony vermilion	Antimonous oxysulfide
Antimony white	Antimonous oxide
Antimony yellow	Basic lead antimonate
Aqua fortis	Nitric acid
Aqua regia	Nitric acid and hydrochloric acid
Argol	Crude potassium acid tartrate
Arsenic glass	Arsenous oxide
Aspirin	Acetyl-salicylic acid
Azurite	Basic copper carbonate
Bakelite	Resin from phenol + formaldehyde
Baking soda	Sodium bicarbonate
Barium white	Barium sulfate
Baryta	Barium oxide
Barytes	Barium sulfate (natural)
Bauxite	Hydrated alumina
Beet sugar	Sucrose
Bentonite	Impure aluminum silicate
Benzene	Mixture of low boiling liquid alkanes
Benzol	Benzene
Bichrome	Potassium dichromate
Bitter salt	Magnesium sulfate
Black ash	Impure sodium carbonate
Blanc-fixe	Barium sulfate (artificial)
Bleaching powder	Calcium chloro-hypochlorite
Blende	Natural zinc sulfide
Blue copperas	Copper sulfate

* This table reproduced from "Handbook of Chemistry and Physics," through the courtesy of Robert C. Weast.

COMMON AND CHEMICAL NAMES OF COMPOUNDS—(Continued)

Common Name	Chemical Name
Blue salts...............	Nickel sulfate
Blue stone...............	Copper sulfate
Blue verditer.............	Basic copper carbonate
Blue vitriol..............	Copper sulfate
Bone ash.................	Impure calcium phosphate
Bone black...............	Crude animal charcoal
Boracic acid..............	Boric acid
Borax....................	Sodium tetraborate
Bremen blue	Basic copper carbonate
Brimstone................	Sulfur
Burnt alum..............	Anhydrous potassium aluminum sulfate
Burnt lime...............	Calcium oxide
Burnt ocher............. } Burnt ore................ }	Ferric oxide
"Butter of"...............	Refers to the chloride
Cadmium yellow..........	Cadmium sulfide
Calamine.................	Zinc silicate
Calcite..................	Mineral calcium carbonate
Caliche..................	Impure sodium nitrate
Calomel..................	Mercurous chloride
Camphor, artificial.........	Pinene hydrochloride
Cane sugar...............	Sucrose
Carbolic acid..............	Phenol
Carbonic acid............ } Carbonic anhydride........ }	Carbon dioxide
Carborundum†.............	Silicon carbide
Carnallite................	Magnesium potassium chloride
"Caustic".................	Refers to the hydroxide of a metal
Ceruse...................	Basic lead carbonate
Chalk....................	Calcium carbonate
Chili niter.............. } Chili saltpeter............ }	Sodium nitrate
China clay...............	Aluminum silicate
Chinese red..............	Basic lead chromate
Chinese white.............	Zinc oxide
Chloramine T.............	Sodium p-toluene-sulfochloramide
Chloride of lime...........	Calcium chloro-hypochlorite
Chloride of soda...........	Sodium hypochlorite solution
Chrome alum.............	Potassium chromium sulfate
Chrome green.............	Chromium oxide
Chrome red..............	Basic lead chromate
Chrome yellow............	Lead chromate
Chromic acid.............	Chromium trioxide
Cinnabar.................	Mercuric sulfide

† Trade name.

221

COMMON AND CHEMICAL NAMES OF COMPOUNDS—(Continued)

Common Name	Chemical Name
Cobalt black...............	Cobalt oxide
Cobalt green..............	Cobalt zincate
Common salt..............	Sodium chloride
Copperas.................	Ferrous sulfate
Corn sugar...............	Glucose
Corrosive sublimate........	Mercuric chloride
Corundum................	Aluminum oxide
Cream of tartar...........	Potassium hydrogen tartrate
Cresylic acid.............	Mixture of *o*-, *m*-, and *p*-cresol
Cupferron................	Nitrosophenylhydroxylamine
Dekaline.................	Decahydronaphthalene
Derby red................	Basic lead chromate
Derinatol................	Basic bismuth gallate
Dextrose.................	Glucose
Dutch liquid..............	Ethylene chloride
Eau-de-Javelle............	Potassium hypochlorite solution
Eau-de-Labarraque........	Sodium hypochlorite solution
Emerald green............	Copper aceto-arsenite
Emery powder............	Impure aluminum oxide
Epsom salts..............	Magnesium sulfate
Essence of bitter almonds.....	Benzaldehyde
Essence of mirbane.........	Nitrobenzene
Everitt's salt.............	Potassium ferrous ferrocyanide
Feldspar.................	Potassium aluminum silicate
Ferro prussiate............	Potassium ferrocyanide
Fixed white..............	Barium sulfate
Flowers of sulfur..........	Sulfur
"Flowers of" a metal.......	A synonym for the oxide
Fluorspar................	Calcium fluoride
Formalin.................	40% solution of formaldehyde in water
Formin..................	Hexamethylene tetramine
Freezing salt.............	Crude sodium chloride
French chalk.............	Hydrated silicate of magnesium
French verdigris..........	Basic copper acetate
Fruit sugar..............	Fructose
Fuller's earth............	Hydrated magnesium and aluminum silicates
Fulminate of mercury.......	Mercuric fulminate
Fusel oil................	Mixed amyl alcohols
Gasoline................	Mixture of low boiling hydrocarbons suitable for use in internal combustion engines
Galena..................	Natural lead sulfide
Glauber's salt............	Sodium sulfate
Glucose.................	Dextrose
Glycerin.................	Glycerol
Grain alcohol............	Ethyl alcohol

Common Name	Chemical Name
Grape sugar	Glucose
Green verditer	Basic copper carbonate
Green vitriol	Ferrous sulfate
Gypsum	Calcium sulfate
Hartshorn salt	Ammonium carbonate carbamate
Heavy spar	Barium sulfate
Hexamine	Hexamethylene tetramine
Horn silver	Silver chloride
Hypo	Sodium thiosulfate
Indian red	Ferric oxide
Iron black	Precipitated antimony
Iron mordant	Ferric sulfate
Kainit	Double salt of potassium magnesium sulfate and magnesium chloride
Kaolin	Aluminum silicate
Kieselguhr	Siliceous earth
Kieserite	Mineral magnesium sulfate
King's yellow	Arsenous sulfide
Lampblack	Impure carbon
Lanolin	Mixture of cholesterol and esters
Laughing gas	Nitrous oxide
Lemon chrome	Barium chromate
Levulose	Fructose
Lime	Calcium oxide
Litharge	Lead monoxide
Lithopone	Zinc sulfide + barium sulfate
Liver of sulfur	Mixed potassium sulfides
Lunar caustic	Silver nitrate
Lysol	Cresol soap solution
Magnesia	Magnesium oxide
Magnesite	Magnesium carbonate
Malachite	Basic copper carbonate
Manganese black	Manganese dioxide
Marble	Calcium carbonate
Marsh gas	Methane
Massicot	Lead monoxide
Methanol	Methyl alcohol
Metol	p-Methylaminophenol sulfate
Microcosmic salt	Sodium ammonium hydrogen phosphate
Milk sugar	Lactose
Milk of barium	Barium hydroxide
Milk of lime	Calcium hydroxide
Milk of magnesium	Magnesium hydroxide
Milk of sulfur	Precipitated sulfur
Minium	Lead tetroxide

COMMON AND CHEMICAL NAMES OF COMPOUNDS—(Continued)

Common Name	Chemical Name
Mohr's salt	Ferrous ammonium sulfate
Molybdenite	Molybdenum disulfide
"Muriate of" a metal	Chloride of the metal
Muriatic acid	Hydrochloric acid
Naphtha (Petroleum)	A petroleum distillate
Naphtha (Solvent)	A coal tar distillate
Natron	Sodium carbonate
Niter	Potassium nitrate
Nitro-lime	Calcium cyanamide
Nitrous ether	Ethyl nitrite
Nordhausen acid	Fuming sulfuric acid
Oil of bitter almond	Benzaldehyde
Oil of garlic	Allyl sulfide
Oil of mirbane	Nitrobenzene
Oil of mustard, artificial	Allyl isothiocyanate
Oil of pears	Amyl acetate
Oil of pineapple	Ethyl butyrate
Oil of vitriol	Concentrated sulfuric acid
Oil of wintergreen, artificial	Methyl salicylate
Oleum	Fuming sulfuric acid
Olifiant gas	Ethylene
Orpiment	Arsenic trisulfide
Osmic acid	Osmic tetroxide
Paris blue	Ferric ferrocyanide
Paris green	Copper aceto-arsenite
Pearl ash	Potassium carbonate
Permanent white	Barium sulfate
Petroleum ether	Mixture of hydrocarbons boiling from 40 to 60°C.
Phenic acid	Phenol
Phosgene	Carbonyl chloride
Phosphate rock	Calcium phosphate
Picric acid	*sym*-Trinitrophenol
Plaster of paris	Calcium sulfate
Plumbago	Graphite
Precipitated chalk	Calcium carbonate
Prussian blue	Ferric ferrocyanide
Prussic acid	Hydrocyanic acid
Putty powder	Impure stannic oxide
Pyrites	Ferrous di-sulfide
Pyroligneous acid	Crude acetic acid
Pyroligneous spirit	Methyl alcohol
Pyrolusite	Manganese dioxide
Quicklime	Calcium oxide
Quicksilver	Mercury
Quinol	Hydroquinone

Common Name	Chemical Name
Realgar	Arsenic disulfide
Rectified spirit	Alcohol 90–5%
Red antimony	Antimony oxysulfide
Red lead	Lead tetroxide
Red liquor	Aluminum acetate solution
Red precipitate	Oxide of mercury
Red prussiate of potash	Potassium ferricyanide
Rochelle salt	Potassium sodium tartrate
Rock salt	Sodium chloride
Rouge	Ferric oxide
Saccharin	Benzoic sulfimide
Sal ammoniac	Ammonium chloride
Salol	Phenylsalicylate
Salt	Sodium chloride
Salt cake	Impure sodium sulfate
Salt of amber	Succinic acid
Salt of lemon / Salt of sorrel	Potassium acid oxalate
Salt of tartar / Salt of wormwood	Potassium carbonate
Saltpeter	Potassium nitrate
Salvarsan	3, 3'-Diamino-4, 4'-dihydroxy-arsenobenzene dihydrochloride
Satin white	Calcium sulfate
Scheele's green	Copper hydrogen arsenite
Schlippe's salt	Sodium thioantimonate
Silica	Silicon dioxide
Slaked lime	Calcium hydroxide
Soda, washing	Sodium carbonate
Soda crystals	Sodium carbonate
Soda lime	Mixture of calcium oxide and sodium hydroxide
Sodium hyposulfite	Sodium thiosulfate
Soft soap	Potash soap
Soluble glass	Sodium silicate
Soluble tartar	Potassium tartrate
Spirit of hartshorn	Ammonia solution
Spirit of salt	Hydrochloric acid
Spirit of wine	Ethyl alcohol
Stassfurtite	Magnesium borate and chloride double salt
Sugar of lead	Lead acetate
Sugar of milk	Lactose
Sulfuric ether	Diethyl ether
Superphosphate	Impure calcium acid phosphate
Sylvine	Potassium chloride
Sylvinite	Sylvine with rock salt

COMMON AND CHEMICAL NAMES OF COMPOUNDS—(Continued)

Common Name	Chemical Name
Table salt	Sodium chloride
Talc	Hydrated magnesium silicate
Tartar	Crude potassium bitartrate
Tartar emetic	Potassium antimonyl tartrate
Tetralin	Tetrahydronaphthalene
Tin crystals	Stannous chloride
Tin white	Stannic hydroxide
T.N.T.	Trinitrotoluene
Toluol	Toluene
Trona	Natural sodium carbonate
Turnbull's blue	Ferrous ferricyanide
Ultramarine yellow	Barium chromate
Unslaked lime	Calcium oxide
Vanillin	Methyl ether of protocatechualdehyde
Venetian red	Ferric oxide
Verdigris	Basic copper acetate
Vermilion	Red mercuric sulfide
Vitriol	Sulfuric acid
"Vitriolate of"	"Sulfate of"
Washing soda	Sodium carbonate
Water glass	Sodium silicates dissolved in water
White acid	Hydrofluoric acid and ammonium fluoride
White arsenic	Arsenous oxide
White lead	Basic lead carbonate
White vitriol	Zinc sulfate
Whiting	Calcium carbonate
Witherite	Barium carbonate
Wood alcohol	
Wood naphtha	Methyl alcohol
Wood spirit	
Xylol	Xylene
Yellow prussiate of potash	Potassium ferrocyanide
Zinc blende	Mineral zinc sulfide
Zinc vitriol	Zinc sulfate
Zinc white	Zinc oxide

Pigments named in the above list refer to the pure substance and not to mixtures often sold under the same name.

INDEX